[英国]埃里克·塞利 著　杨晨 译

元素周期表

牛津通识读本·

The Periodic Table

A Very Short Introduction

译林出版社

图书在版编目（CIP）数据

元素周期表／（英）埃里克·塞利（Eric Scerri）著；杨晨译．—南京：译林出版社，2022.8（2024.2重印）
（牛津通识读本）
书名原文：The Periodic Table: A Very Short Introduction
ISBN 978-7-5447-9176-2

Ⅰ.①元… Ⅱ.①埃… ②杨… Ⅲ.①化学元素周期表－普及读物 Ⅳ.①O6-64

中国版本图书馆 CIP 数据核字（2022）第 090960 号

著作权合同登记号　图字：10-2018-120 号

元素周期表　［英国］埃里克·塞利／著　杨　晨／译

责任编辑　许　昆
装帧设计　景秋萍
校　　对　戴小娥
责任印制　董　虎

原文出版　Oxford University Press, 2019
出版发行　译林出版社
地　　址　南京市湖南路 1 号 A 楼
邮　　箱　yilin@yilin.com
网　　址　www.yilin.com
市场热线　025-86633278
排　　版　南京展望文化发展有限公司
印　　刷　江苏扬中印刷有限公司
开　　本　890 毫米 ×1260 毫米　1/32
印　　张　10
插　　页　4
版　　次　2022 年 8 月第 1 版
印　　次　2024 年 2 月第 2 次印刷
书　　号　ISBN 978-7-5447-9176-2
定　　价　39.00 元

序　言

郭子建

元素周期表是化学科学的基石，在化学发展史上具有里程碑意义。元素周期表将物质的基本成分有序组织起来，反映出元素的电子结构、化学性质和反应行为，毫不夸张地说，几乎所有化学知识均起始于元素周期表。

介绍元素周期表的书籍非常多，但《元素周期表》这本科普读物更加值得一读。此书用简短的篇幅、通俗易懂的语言总结了元素周期表的渊源、意义以及与元素相关的最新发现，可读性很强。

作为元素周期表中的“住户”，元素的发现几乎与人类文明相伴。像铁、铜、金、银等容易以单质形态存在的元素发现得最早，之后从青铜时代到铁器时代，从炼金术到电解技术，从核辐射到核裂变，元素周期表逐步被填充并扩展，新“住户”不断占据各自的位置。直至今日，仍然有新的元素被发现（合成），而且关于元素在生命起源、物种进化中的作用以及元素如何精准调控材料性能及生命过程等仍然是科学家们非常关注的科学问题。

元素以及元素周期系统的发现体现了科学家的智慧和执着精神，非常具有故事性。大家对拉瓦锡、道尔顿、洪堡、盖-吕萨克、阿伏伽德罗、门捷列夫、居里夫人等科学家发现元素、总结元素周期规律的故事耳熟能详，但也有像法国地质学家尚古尔多阿这样做出了关键发现而被忽视的科学家的例子。现代元素周期表经历了150多年的修改和完善，进化出众多外形各异、种类繁多的展示方式，但门捷列夫等先驱所“搭建”的元素周期系统得到了完美保留，成为成果最丰富、结构最统一的科学体系之一。元素在周期表不同行列中排列的方式，能反映出它们的许多关系，为解释和预测元素的理化性质及可能的价键结构奠定了基础。目前大家在教科书中，在实验室和教室等场所看到最多的是中长式表，该表将镧系和锕系元素排在主表的下方。

与元素周期表排布方式密不可分的重要问题是为什么元素会出现周期性重复但又间隔不同的变化规律，即周期系统。尽管至今我们仍然无法完美解释，但不可否认，量子力学通过推断电子在核周围位置的排列推动了研究周期系统的进程，为解析周期系统提供了巨大帮助。波尔将量子力学应用在原子体系中，提出周期表中任意一族元素之间的相似性均在于它们拥有数量相等的外层电子。之后人们将电子看作粒子同时又看作波，电子不再沿确定的轨迹或者说轨道绕原子核运动，而用弥散的电子云来描述。以周期系统为指导，即使人们不了解100多种元素的性质，仍然可以根据它在周期表中所处的位置（主族、过渡金属、镧系、锕系）以及这些位置上下典型已知“住户”的性质对其做出有效预测。钇钡铜氧超导体的探索、非铂类抗肿瘤药物的发现等均是基于同族元素化学相似性而展开的前沿研究。

元素周期表的发现非常伟大，因此，联合国教科文组织为纪念门捷列夫发现元素周期表150年，宣布2019年为“国际化学元素周期表年”。我希望与阅读此书的读者朋友分享的一点是，元素的发现以及元素周期表的完善永无止境，能更加完美地诠释元素性质的周期表一定会出现。

目 录

致谢与献词

感谢牛津大学出版社的全体编辑和其他工作人员，包括提出“牛津通识读本”计划的杰瑞米·路易斯。我还要感谢在本书准备期间提供了帮助的各位同事、学生和图书管理员。

本书献给我的妻子埃莉莎·塞德纳。

前 言

关于周期表的奇妙，人们已经写了很多。这里仅举几例。

周期表是大自然的罗塞塔石碑。对不了解它的人而言，它只是100多个编了号的盒子，每个包含一两个字母，以一种奇怪的对称方式排列。然而对化学家而言，周期表揭示了物质的组织原理，也就是化学的组织原理。从根本上说，化学的全部内容都包含在周期表当中。

当然，这不是说化学的全部内容可以明显从周期表中看出来，绝非如此。但周期表的结构反映了元素的电子结构，因此反映了元素的化学性质和行为。也许更恰当的说法是，化学的全部内容都始于周期表。

（鲁迪·鲍姆，《化学化工新闻·元素特刊》）

天文学家哈洛·沙普利写道：

周期表或许是人类迄今为止设计出的最紧凑、最有意义的知识汇编。周期表对物质的作用与地质年代表对宇宙时的作用一样。它的历史讲述了人类在微观宇宙中的伟大征服。

化学史学家罗伯特·希克斯则在网络播客中如是说:

也许在所有科学中,最容易辨识的标志就是元素周期表。这份图表已经成为我们的模型,来说明原子和分子如何排布自己创造我们所知的物质,还有世界如何在最细微的层面上组织起来。纵观历史,元素周期表也在变化。新发现的元素加入表中,另一些元素被发现有误,要么被修改要么被移除。如此,周期表就像一间化学历史的仓库,是化学科学当前发展的模板、未来发展的基础……是世界最基本的构造材料的地图。

现在,以C. P. 斯诺,一位以有关"两种文化"的著作而闻名的物理化学家的话作为最后一个例子。

第一次[了解周期表时],我看到一堆杂乱无章的事实一排排按次序排列。童年时期无机化学所有的杂乱、秘诀和大杂烩,仿佛都在眼前融入了这个系统——就好像一个人站在丛林边,看到它突然变成了一座荷兰花园。

周期表的不同寻常之处在于,它既简单又叫人觉得熟悉,并且它在科学中具有真正的基础地位。简单这点在上面的引文中已有暗示。周期表似乎将所有物质的基本成分都组织起来了。大多数人也都知道它。每一个对化学只有基本了解的人，或许会把学过的一切化学知识都忘了，但几乎都能想起来还有周期表这回事。周期表就像水的化学式一样为人熟知。它已经成为真正的文化标志，被艺术家、广告商，当然还有各类科学家使用。 iv

同时，周期表也不只是化学教学和学习的工具。它反映了世界上各个事物的自然规律，就我们所知，也反映了整个宇宙的自然规律。它由一族族按列排列的元素组成。如果化学家（或哪怕是化学专业的学生）知道任何一族中一种典型元素（比如钠）的性质，他就能很好地知道同族其他元素（如钾、铷、铯）的性质。

更为根本的是，周期表的固有顺序让人们深刻认识了原子结构，认识到电子本质上是在特定的壳层和轨道绕核运动的这一概念。电子的这些排列反过来也使周期表变得合理。大体上说，它们首先解释了钠、钾、铷等元素为什么属于同一族。但更重要的是，因尝试理解周期表而率先得到的对原子结构的理解，已经应用到了科学中的许多其他领域。首先，这些知识促进了旧量子理论的发展，之后又促进了它更成熟的表弟——量子力学的发展。量子力学这套知识接着成为物理学的基本理论，不仅可以解释所有物质的行为，也可以解释所有形式的辐射，如可见光、X射线和紫外线等。

周期表与19世纪大多数科学发现不同，它没有被20、21世纪的发现驳倒。而且，现代物理的发现反而使周期表更完善，并规整了一些遗留的反常问题。但它的整体形式和有效性仍然完 v

好无损，这是这一知识体系所具力量和深度的又一明证。

在研究周期表之前，我们先来看看它的住户——元素。然后从第三章开始进入其历史，看看我们如何达到当前对它的理解程度，最后再快速了解一下现代周期表和它的一些变形。vi

第二版前言

“牛津通识读本”的负责人要求我编写第二版，以配合门捷列夫发表成熟周期表150周年、联合国教科文组织宣布2019年为“国际化学元素周期表年”这一活动。

不用说，我愿意借这次机会纠正读者指出的第一版中的全部错误，并更新材料，因为周期表和相关问题的研究取得了一些新进展。

例如，在周期表的现代表述中很重要的4s-3d轨道占据问题，如今有了一个令人满意的解答，充分解释了这些原子轨道在填充和电离时的相对顺序。此外，对构造原理与其反常现象之间的关联也重新做了概念说明，并会在此新版中给出简要解释。

合成元素领域也有了一些新进展。例如，自第一版面世以来，第113、第115、第117和第118号元素被正式承认并命名为鉨、镆、石田和氮。这一进展也标志着人们发现周期表后首次完成第七周期的全部命名。不过，尝试合成下一周期元素（从第119和第120号元素开始）的工作已经展开。至少在形式上，g区

vii

元素将从第121号元素开始并继续下去。

对于哪些元素应当占据周期表第3族的问题，以及针对氦的位置问题所做的一些新实验，也都有了新的进展。

读者如对本书提出的问题有任何疑问或建议，希望能与我联系。

埃里克·塞利

洛杉矶

2019年

第一章

元 素

古希腊哲学家只确认了土、水、气和火四种元素，它们都还保留在占星术对黄道十二宫的划分当中。[①]那时的一些哲学家认为，不同的元素由不同形状的微小成分构成，这样就解释了这些元素的不同性质。他们设想，四元素的基本形状对应着柏拉图多面体（图1），即由大小相同的三角形、正方形等二维图形拼成的立体图形。他们认为土由微小的立方体颗粒构成。如此联想，是因为在所有柏拉图多面体中，立方体每个面的面积最大。[②]水能流动，可以解释为它具有正二十面体那种更流畅的形状；而人碰到火会感到疼痛，是因为火由正四面体这样尖锐的颗粒构成。气由正八面体构成，因为剩下的柏拉图多面体只有它了。一段时间之后，数学家发现了第五种柏拉图多面体，即正十二

① 占星术将黄道十二宫分为火象星座、土象星座、风象星座和水象星座四类，每一类包含三个星座。在占星术中，“气”一般称作“风”。——译注（若无特别说明，本书注释均为译注）

② 此处的柏拉图多面体只有四种，不包括正十二面体。这四种中只有立方体的面是正方形，其他三种都是正三角形，因此说立方体每个面的面积最大。

面体。这使得亚里士多德提出可能还存在第五种元素，即“精华”，也叫作以太。

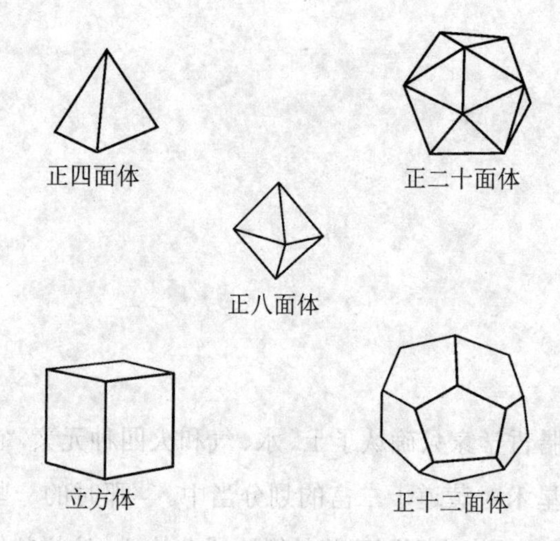

图1　柏拉图多面体，每个对应一种古代元素

如今，元素由柏拉图多面体构成的观念已然被视为谬误，但它是另一个硕果累累的观念的源头：物质的宏观性质由组成它们的微观成分的结构决定。这些“元素”一直很好地流传到了中世纪及之后，当中还添加了一些由炼金术士（现代化学家的先祖）发现的其他元素。炼金术士最为人知的目标就是实现元素转化。他们尤其希望将贱金属铅转变成贵金属金，后者因其色泽、稀有程度和稳定的化学性质，自文明诞生起就是最受珍视的物质之一。

不过，古希腊哲学家除了把“元素”看作实际存在的物质之外，还认为它们是一种原则，也就是能够引出元素物质外在性质的倾向与潜能。这种非常细致地区分元素抽象形式与外在形式的做法，在化学发展中扮演了很重要的角色，尽管当前即便专业化学家也不能很好地理解当中更微妙的含义。然而，抽象元素的概念为周期系统的各位先驱（比如主要发现者德米特里·门捷列夫）提供了基本的指导原则。 2

按大多数教材的说法，化学在抛弃了古希腊知识、炼金术和对元素本质近乎神秘的理解之后，才繁荣发展起来。人们普遍认为，现代科学胜在依靠直接实验，坚持可观测才可信的原则。元素概念中那些更玄妙，或者说更基本的意义渐渐不再被人接受，也是意料之中。例如，安托万·拉瓦锡的观点就是元素必须通过经验观察来定义，这降低了抽象元素或元素作为原则的作用。拉瓦锡认为，元素应当定义为一种尚待分割成更基本成分的物质实体。1789年，拉瓦锡发表了一份包含33种单质（或者按上面的经验标准称为元素）的列表（图2）。这份元素列表正确地抛弃了土、水、气、火四种古代元素：现在已经证明，它们都由更简单的物质构成。

以现代的标准来看，拉瓦锡列表中有许多物质都可以认定是元素。但剩下的如lumière（光）和calorique（热）等就不再是元素了。之后，分离和界定化学物质的技术突飞猛进，为化学家不断扩充这份列表提供了助力。光谱分析是一项测量不同物质的发射和吸收光谱的重要技术，它最终发展为一种非常精确的方法，能够通过元素的“指纹”来确认各种元素。现在，我们已经识别出大约90种自然存在的元素。此外，还人工合成了大概25种元素。 3

	Noms nouveaux.	*Noms anciens correſpondans.*
Subſtances ſimples qui appartiennent aux trois règnes, & qu'on peut regarder comme les élémens des corps.	Lumière	Lumière.
	Calorique........	Chaleur.
		Principe de la chaleur.
		Fluide igné.
		Feu.
		Matière du feu & de la chaleur.
	Oxygène	Air déphlogiſtiqué.
		Air empiréal.
		Air vital.
		Baſe de l'air vital.
	Azote..........	Gaz phlogiſtiqué.
		Mofète.
		Baſe de la mofète.
	Hydrogène.......	Gaz inflammable.
		Baſe du gaz inflammable.
Subſtances ſimples non métalliques oxidables & acidifiables.	Soufre..........	Soufre.
	Phoſphore	Phoſphore.
	Carbone	Charbon pur.
	Radical muriatique .	Inconnu.
	Radical fluorique...	Inconnu.
	Radical boracique. .	Inconnu.
Subſtances ſimples métalliques oxidables & acidifiables.	Antimoine	Antimoine.
	Argent	Argent.
	Arſenic.........	Arſenic.
	Biſmuth	Biſmuth.
	Cobalt	Cobalt.
	Cuivre..........	Cuivre.
	Etain	Etain.
	Fer............	Fer.
	Manganèſe.......	Manganèſe.
	Mercure	Mercure.
	Molybdène	Molybdène.
	Nickel..........	Nickel.
	Or............	Or.
	Platine	Platine.
	Plomb	Plomb.
	Tungſtène.......	Tungſtène.
	Zinc	Zinc.
Subſtances ſimples ſalifiables terreuſes.	Chaux..........	Terre calcaire, chaux.
	Magnéſie	Magnéſie, baſe du ſel d'epſom.
	Baryte	Barote, terre peſante.
	Alumine	Argile, terre de l'alun, baſe de l'alun.
	Silice	Terre ſiliceuſe, terre vitrifiable.

图2　拉瓦锡的单质元素列表

元素的发现

像铁、铜、金、银这些元素，在文明诞生之时就已经为人所知。这说明了这样的事实：这些元素能够以单质形态出现，或者很容易从它们所在的矿石中分离出来。

历史学家和考古学家把人类历史上的某些时期称为铁器时代和青铜时代（青铜是铜与锡的合金）。炼金术士又向元素列表中添加了一些元素，比如硫、汞、磷等。在相对现代的时期，电的发现使化学家可以分离出许多性质更加活泼的元素——这些元素不像铜和铁，没法用矿石和木炭（碳）一起加热得到。化学史上有好几段重要的时期，研究者在几年内就能发现六七种元素。比如，英国化学家汉弗莱·戴维用电（或者更准确地说，用电解技术）分离出了十种元素，包括钙、钡、镁、钠、氯等。

人们发现放射性与核裂变之后，又发现了更多元素。自然存在条件下，最后分离出来的七种元素分别是镤、铪、铼、锝、钫、砹和钷，分离时间在1917年到1945年之间。最后填上的空缺是第43号元素，也就是锝。它得名于希腊语的techne，意思是“人造的”。它由辐射化学反应过程“制造”出来，这在核物理出现之前不可能做到。不过，现在我们知道，锝在地球上确实能够自然存在，只是含量非常低而已。

元素的命名

元素周期表的吸引力，一部分来源于元素的独特性质，比如它们的颜色、触感等。还有很大一部分则在于它们的名字。化学家、集中营幸存者普里莫·莱维写过一本广受好评的书，就叫

《元素周期表》，它的每一章都以一种元素命名。这本书主要讲述了莱维与亲人、熟人的故事，每段故事都因他对某种元素的喜爱而起。神经学家、作家奥利弗·萨克斯写过一本名为《钨舅舅》的书，讲述了他对元素、化学，尤其是对周期表的痴迷。最近还有萨姆·基恩和修·奥尔德西-威廉姆斯写的两本关于元素的畅销书。我想可以这么说，元素吸引公众想象力的时代现在真的到来了。

发现元素后的几百年来，研究者用了许多不同的方式来给元素命名。第61号元素钷的名字（promethium）来源于普罗米修斯（Prometheus）。这位天神从天界将火盗予人类，因此遭受宙斯惩罚。这个故事和第61号元素的关联是，研究者在分离它时付出了很大努力，就像希腊神话中的普罗米修斯那般英勇，不畏艰险。钷是地球上少数非自然存在的元素之一，它是另一种元素——铀——经过核裂变的衰变产物。

行星和其他天体也被用于命名元素。氦元素的名字（helium）来自helios，希腊语中的太阳。它最早是1868年从太阳的光谱中发现的，直到1895年才从地球上的物质中识别出来。类似的还有以小行星智神星（Pallas）命名的钯（palladium），而智神星的名字又来自希腊的智慧女神帕拉斯（Pallas）[①]。元素铈（cerium）以人类1801年发现的第一颗小行星谷神星（Ceres）命名。铀（uranium）以天王星（Uranus）命名，两者都发现于18世纪80年代。许多这类例子都有神话的主题。例如，乌拉诺斯（Uranus）就是希腊神话中的天空之神。

① 即雅典娜。

许多元素因其颜色而得名。黄绿色的气体氯（chlorine）得名于希腊语khloros，指的就是黄绿色。铯（caesium）得名于拉丁语cesium，意思是蓝灰色，因为它的光谱中有明显的蓝灰线。元素铑（rhodium）的盐是粉红色的，所以研究者为它选了rhodon这个词，即希腊语里的蔷薇。金属铊（thallium）得名于拉丁语thallus，意思是绿枝。这种元素是英国化学家威廉·克鲁克斯因其光谱中有明显的绿线而发现的。

很大一部分元素的名字来源于发现者所居住的或希望纪念的地方。比如镅、锫、锎、𫟼、铕、钫、锗、𬭶、钋、镓、铪（hafnium，出自哥本哈根的拉丁名Hafnia）、镥（lutetium，出自巴黎的拉丁名Lutetia）、镆、鉨（nihonium，出自日本的罗马字Nihon）、铼（rhenium，出自莱茵河流域的英文名Rhine）、钌（ruthenium，出自Rus，拉丁文中指今天的俄罗斯西部、乌克兰、白俄罗斯、斯洛伐克的一部分和波兰等地区）和石田。[①] 还有的元素名得自发现它们的矿场的地理位置。这一类包括四种以瑞典村庄伊特比（Ytterby，靠近斯德哥尔摩）命名的元素：铒（erbium）、铽（terbium）、镱（ytterbium）和钇（yttrium），它们都是在村庄附近的一处矿区发现的。还有第五种元素钬（holmium），以斯德哥尔摩的拉丁名Holmensis命名。

对于更近期合成的元素，它们的名字多来自发现者或发现者希望纪念的人。例如，𬭳、锔、锿、镄、𫓧、铹、𬬻、钔、锘、鿬、轮、

① 其中因英文较明显故作者没有解释的，补充解释如下：镅（americium）出自美国，锫（berkelium）出自伯克利，锎（californium）出自加利福尼亚州，𫟼（darmstadtium）出自重离子研究所所在的达姆施塔特市，铕（europium）出自欧洲，钫（francium）出自法国，锗（germanium）出自德国，𬭶（hassium）出自达姆施塔特所在的黑森州，钋（polonium）出自波兰，镓（gallium）出自高卢，镆（moscovium）出自莫斯科，石田（tennessine）出自田纳西州。

铲、镭都是如此。[①]

表尾的超铀元素命名带有民族主义争端的特征，在一些情况下，研究者对是谁第一个合成了某元素及由此在选名上应该纪念谁，会爆发相当激烈的争论。为了试着解决这类争端，国际纯粹与应用化学联合会规定，在任何情况下，元素都应当公平且系统地用原子序数的拉丁数字命名。比如，第105号元素就是un-nil-pentium（Unp），而第106号元素就是un-nil-hexium（Unh）。不过最近，国际纯粹与应用化学联合会在审议了一些最新的超重元素后，又将命名权还给了在各自情况下确定了优先权的发现者或合成者。第105和第106号元素现在分别称作钍（dubnium）和镭（seaborgium）。

周期表中用来标记各个元素的符号也有丰富有趣的故事。在炼金术时代，元素符号通常和它们从之得名或与之相关的天体的符号一致（图3）。例如，汞元素就和水星（太阳系最靠内的行星）共用一个符号。铜和金星相关，二者也共用一个符号。

约翰·道尔顿在1805年发表他的原子理论时，就为元素保留了几个炼金术符号。然而，这些符号非常不方便，在文章和书籍里也很难复制。使用字母符号的现代方法是瑞典化学家琼斯·雅可比·贝采尼乌斯在1813年开始采用的。

现代周期表里有很少一部分元素用单个字母表示，比如氢、碳、氧、氮、硫、氟等，分别用H、C、O、N、S、F等表示。大部分元素用两个字母表示，其中首字母大写，另一个小写。例如，我们分

① 对应如下：铍（玻尔）、锔（居里夫人）、锿（爱因斯坦）、镄（费米）、铁（苏联原子物理学家弗洛伊洛夫）、铹（劳伦斯）、鿏（奥地利裔瑞典原子物理学家迈特纳）、钔（门捷列夫）、锘（诺贝尔）、鿫（苏联物理学家奥加涅）、轮（伦琴）、铲（卢瑟福）、镭（美国核物理学家西博格）。

金属	金	银	铁	汞	锡	铜	铅
符号	☉	☾	♂	☿	♃	♀	♄
天体	Sun（太阳）	Moon（月球）	Mars（火星）	Mercury（水星）	Jupiter（木星）	Venus（金星）	Saturn（土星）
日							
拉丁文	*Solis*	*Lunae*	*Martis*	*Mercurii*	*Jovis*	*Veneris*	*Saturni*
法文	*dimanche*	*lundi*	*mardi*	*mercredi*	*jeudi*	*vendredi*	*samedi*
英文	Sunday	Monday	Tuesday	Wednesday	Thursday	Friday	Saturday

图3　古代元素的名称和符号

别用Kr、Mg、Ne、Ba、Sc来表示氪、镁、氖、钡、钪。有些双字母符号直观上完全看不出来源，比如Cu、Na、Fe、Pb、Hg、Ag、Au等，它们来自铜、钠、铁、铅、汞、银、金的拉丁名。钨用W表示，出自该元素的德语名wolfram。

第二章

现代周期表速览

现代周期表

元素在周期表行列中排列的方式，能反映出它们的许多关系。其中一些关系广为人知，另一些则依然有待发现。在20世纪80年代，科学家发现超导性（即电在零电阻下流动）出现的温度比之前观测到的超导温度高出许多，从不高于20 K的典型温度值跃升到100 K级别。

研究者在将镧、铜、氧、钡结合成复杂化合物时，偶然发现它具有高温超导性。随后，全世界争先恐后展开行动，用尽办法提高能够维持超导效应的温度。最终的目的是做到常温超导，它可以带来许多技术突破，比如让磁悬浮列车在超导轨道上轻松滑行。元素周期表是这项探索用到的一条主要原理。它可以让研究者将化合物中的某些元素替换成行为类似的其他元素，再测试产物的超导性质。元素钇就是这么进入了一组新的超导化合物：化合物$YBa_2Cu_3O_7$的超导温度为93 K。这些知识，当然还有更多的知识，就潜藏在周期系统中，等待着被发现、被善用的那天。

就在最近，研究者又发现了一类新的高温超导体。它们是氮磷族氧化物，一类包含氧元素、一种氮族（第15族）元素，以及一种或多种其他元素的材料。LaOFeP和LaOFeAs的超导性质（分别发现于2006年和2008年）发表后，研究者对这类化合物的兴趣暴增。他们再次想到了使用砷（As）（就像上文的后一种化合物那样），它在周期表里就位于磷的正下方。

制药领域对同族元素的化学相似性也怀有很大的兴趣。例如，铍元素位于周期表第2族顶端，在镁之上。因为两种元素相似，铍能替代药物中的镁元素，而镁对人体必不可少。这种行为从一个方面解释了铍为什么对人体有毒，因为它们固然很相似，但终究是不同的。类似的还有镉元素，它在周期表中位于锌的正下方。于是，在许多重要的酶中，它能替换锌。周期表同一行位置相邻的元素也会具有相似的性质。例如，铂与金相邻，而我们很早就知道，一种名为顺铂的含铂无机化合物能够治疗多种类型的癌症。于是，很多药物在研制时就用金替换铂，人们用这种方法已经生产了一些很有效的新药物。

最后再提一个根据元素在周期表中的位置来制药的例子。这种药使用了铷元素，铷在表中属于第1族、位于钾的正下方。像之前提过的一样，铷原子近似钾原子，因此像钾一样容易被人体吸收。研究者将这一行为用于监测技术，因为铷很容易受癌变组织，尤其是脑部癌变组织吸引。

常规周期表按行和列排布。横向纵向都可以观察到元素性质的变化趋势。每一横排都是表上的一个周期。要横跨一个周期，首先会走过左边的钾、钙之类的金属，再穿越铁、钴和镍这类过渡金属，然后经过锗这样的半金属元素，再继续走到表右诸如砷、硒和溴等非金属那里。通常，在横跨整个周期时，元素的化学、物理性质都会连续变化，但这一通则也有例外，化学研究因而成为一个引人入胜又难以预料的复杂领域。

金属多样，可以从柔软暗淡如钠、钾渐变到坚硬闪亮如铬、铂和铁。另一方面，非金属往往是固态和气态的，分别如碳和氧那样。就外观而言，固态金属和固态非金属常常很难分辨。在非专业人士看来，坚硬闪亮的非金属可能比钠这样的软金属更像金属。从金属到非金属的周期变化在每个周期循环出现，因此把每行摞起来之后，每列（或者说每族）就由相似的元素构成。每一族内部的元素往往共有很多重要的理化性质，尽管也有不少例外。

1990年，国际纯粹与应用化学联合会推荐用阿拉伯数字取代罗马数字，为各族元素从左到右依序编为第1到第18族（图4），并不再使用旧周期表中的字母A、B等。

H																	He
Li	Be											B	C	N	O	F	Ne
Na	Mg											Al	Si	P	S	Cl	Ar
K	Ca	Sc	Ti	V	Cr	Mn	Fe	Co	Ni	Cu	Zn	Ga	Ge	As	Se	Br	Kr
Rb	Sr	Y	Zr	Nb	Mo	Tc	Ru	Rh	Pd	Ag	Cd	In	Sn	Sb	Te	I	Xe
Cs	Ba	Lu	Hf	Ta	W	Re	Os	Ir	Pt	Au	Hg	Tl	Pb	Bi	Po	At	Rn
Fr	Ra	Lr	Rf	Db	Sg	Bh	Hs	Mt	Ds	Rg	Cn	Nh	Fl	Mc	Lv	Ts	Og

La	Ce	Pr	Nd	Pm	Sm	Eu	Gd	Tb	Dy	Ho	Er	Tm	Yb
Ac	Th	Pa	U	Np	Pu	Am	Cm	Bk	Cf	Es	Fm	Md	No

图4　中长式表

周期表的形式

毫不夸张地说，光是已出版的周期表就超过1 000种，近年来网上的就更多了。它们都有什么联系？是否存在最佳周期表？这些是本书即将探讨的问题，它们可以教给我们许多现代科学中的有趣事物。

我们应当首先解决问题中的一点。对已出版的周期表进行分类的一种方法，是将它们分为三种基本形式。第一种是由纽兰兹、洛塔尔·迈耶尔和门捷列夫等周期表先驱最初提出、发表的短式表，我们会适时更细致地讨论它们（图5）。

这类表本质上是将当时已知的所有元素堆为8列，或者说8族。粗略地说，如果元素按自然顺序（这是后面要讨论的另一个话题）排列，那么这种排列就反映了元素似乎会每8个重复一次的事实。随着有关元素性质的信息不断积累，加之又发现了更多元素，一种叫作中长式表（图4）的新型排列脱颖而出。现在，这种表型基本上算是独占鳌头。它有一个奇怪的特点，即表的主体没有包含所有的元素。看看图4，你就会发现第56和第71号元素之间断开了，后面第88和第103号元素之间也一样。“缺失”的元素都堆放到一处，像独立的脚注似的列于主表下方。

将镧系与锕系元素分离出来的做法纯粹是为了方便。如果不拿出来，周期表就会宽很多，准确地说是32个元素宽度，而现在是18个元素宽度。32个元素宽的表不适合印在化学课本的内页，或是做成大幅挂图挂在教室和实验室里。不过，如果将元素用这种扩展的形式排列（有时候会这样做），就得到了长式周期表。从元素顺序不中断的角度来说，这种表比我们熟悉的中

MENDELÉEFF'S TABLE I.—1871.

Series.	GROUP I. R_2O.	GROUP II. RO.	GROUP III. R_2O_3.	GROUP IV. RH_4. RO_2.	GROUP V. RH_3. R_2O_5.	GROUP VI. RH_2. RO_3.	GROUP VII. RH. R_2O_7.	GROUP VIII. RO_4.
1	H=1							
2	Li=7	Be=9.4	B=11	C=12	N=14	O=16	F=19	
3	Na=23	Mg=24	Al=27.3	Si=28	P=31	S=32	Cl=35.5	
4	K=39	Ca=40	—=44	Ti=48	V=51	Cr=52	Mn=55	Fe=56, Ce=59 Ni=59, Cu=63
5	(Cu=63)	Zn=65	—=68	—=72	As=75	Se=78	Br=80	
6	Rb=85	Sr=87	? Y=88	Zr=90	Nb=94	Mo=96	—=100	Ru=194, Rh=104 Pd=106, Ag=108
7	(Ag=108)	Cd=112	In=113	Sn=118	Sb=122	Te=125	I=127	
8	Cs=133	Ba=137	? Di=138	? Ce=140	....	....	....	....
9	....	....	....	....	....	....	....	....
10	....	....	? Er=178	? La=180	Ta=182	W=184	....	Os=195, In=197 Pt=198, Au=199
11	(Au=199)	Hg=200	Tl=204	Pb=207	Bi=208	....	....	
12	....	....	....	Th=231	....	U=240	....	....

图5　门捷列夫1871年发表的短式周期表

长式表更正确。

不过周期表里的“住户”到底是什么？我们且回到常规的周期表，并选择熟悉的中长式表，以此为这个二维网格的骨架（或者说框架）附上些血肉。元素是怎么发现的？元素是什么样的？当我们沿着周期表的一列向下，或横跨一个周期时，元素会怎样变化？

周期表中几族典型的元素

在表的最左端，第1族包含了金属钠、钾、铷等元素。它们的质地出奇柔软，性质非常活泼，与铁、铬、金、银这些我们通常认识中的金属很不一样。第1族的金属非常活泼，只要将一小块投入水中，就会引发剧烈的反应，产生氢气并留下无色的碱性溶液。第2族的元素包括镁、钙、钡等，它们在大多数情况下都不如第1族的元素活泼。

视线向右移动，就看到居中的一块长方形元素群。它们叫作过渡金属，包括铬、镍、铜、铁等元素。在早期的周期表，即短式表中（图5），这些元素都位于我们现在称为主族元素的族里。

在现代周期表中，这些元素与表的主体分开，尽管这么安排的好处要大过坏处，但它们的很多有价值的化学特征都看不到了。在中长式表的过渡金属右侧，是另一块代表元素，从第13族开始，到第18族结束。表的最右边是惰性气体。

有时，同族的共有性质并不特别明显。第14族就是这种情况，它由碳、硅、锗、锡、铅等构成。沿这一族向下，你会看到非常大的变化。族首的碳是非金属固体，有三种完全不同的结构形式（金刚石、石墨和富勒烯），它也是构成所有生命体的基础。

下一个元素硅是半金属，有意思的是，它构成了人工生命，或至少是人工智能的基础，因为它位于所有计算机的核心。下一个元素锗是半金属，发现时间要晚得多。门捷列夫曾预言锗的存15 在，之后人们发现他预测的很多性质都是对的。往下来到锡和铅，这是两种古代就已知的金属。尽管第14族的元素从金属-非金属行为的角度看有巨大的差异，但它们仍然在一项重要的化学意义上是相似的：它们的最大化合价都是4，也就是都能形成四条键。

第17族元素甚至有着更显著的变化。这一族打头的氟元素和氯元素都是有毒气体。下一个成员溴是已知仅有的两种室温下为液体的元素之一（另一种是金属汞）。继续往下走，你会遇到碘，它是一种紫黑色的固体元素。如果让一名化学新手按照元素的外观来分组，他恐怕不太可能将氟、氯、溴、碘分在一组。在这种情况下，元素概念的感观意义与抽象意义之间的细微差别就起作用了。它们之间的相似性，主要在于抽象元素的性质，而不在于可以分离出来并单独观察的实体特点，对此弗里茨·帕内特做过相关讨论。（见“延伸阅读”）

一路向右，会遇到一族不寻常的元素。它们是惰性气体，都在19、20世纪之交才首次被分离出来。颇为矛盾的是，至少在最初分离出来之时，它们的主要性质就是几乎没有化学性质。这些元素包括氦、氖、氩、氪等，早期的周期表里甚至都没有它们，因为人们不知道，而且完全想不到它们存在。它们被发现后，其存在对周期系统提出了严峻的挑战，但研究者扩展了周期表，增17 加了标为第18族的一个族，最终成功接纳了它们。

你在现代周期表底部看到的那一块元素由镧系与锕系元素

组成，通常画成与主体不连续的形式。但它只是常规周期表的一个显眼的布局特点。就像过渡金属一般以一整块插在表的主体中部，镧系与锕系元素当然也可以这样。而且，我们已经出版了许多这样的长式表。虽然长式表（图6）为这些元素安排了一个更自然的位置，让它们和其他元素排在一起，但这种表太麻烦，不适合做成方便张贴的周期系统挂图。尽管周期表各式各样，种类繁多，但不论其呈现形式如何，整个构架的背后都是周期律。

周期律

周期律说的是，在相当规律但又不断改变的间隔里，化学元素会近似地重复它们的性质。例如，氟、氯、溴处在第17族，它们都有与金属钠结合成化学式为NaX的白色晶体盐的性质（其中X为任意卤族元素）。这种性质上的周期重复是周期系统一切现象背后的本质事实。

周期律这个话题带来了一些有趣的哲学问题。首先，元素的周期性既不恒定也不精准。在常规的中长式周期表中，第一行只有2种元素，第二、第三行有8种，第四、第五行有18种，如此等等。这表明周期以2、8、8、18等变化，和我们从日期、星期，还有音阶中的乐音等看到的周期性非常不一样。在后面这些例子中，周期的长度是固定的。比如7是每星期的天数，也是西方音阶中的乐音数。

然而，在元素间，不仅周期的长度会变化，连周期性本身都不是精确的。周期表中任何一列的元素都不会彼此精确地重现。在这方面，它们的周期性倒有些像音阶。在音阶中，当乐音

H 1																															He 2
Li 3	Be 4																									B 5	C 6	N 7	O 8	F 9	Ne 10
Na 11	Mg 12																									Al 13	Si 14	P 15	S 16	Cl 17	Ar 18
K 19	Ca 20															Sc 21	Ti 22	V 23	Cr 24	Mn 25	Fe 26	Co 27	Ni 28	Cu 29	Zn 30	Ga 31	Ge 32	As 33	Se 34	Br 35	Kr 36
Rb 37	Sr 38															Y 39	Zr 40	Nb 41	Mo 42	Tc 43	Ru 44	Rh 45	Pd 46	Ag 47	Cd 48	In 49	Sn 50	Sb 51	Te 52	I 53	Xe 54
Cs 55	Ba 56	La 57	Ce 58	Pr 59	Nd 60	Pm 61	Sm 62	Eu 63	Gd 64	Tb 65	Dy 66	Ho 67	Er 68	Tm 69	Yb 70	Lu 71	Hf 72	Ta 73	W 74	Re 75	Os 76	Ir 77	Pt 78	Au 79	Hg 80	Tl 81	Pb 82	Bi 83	Po 84	At 85	Rn 86
Fr 87	Ra 88	Ac 89	Th 90	Pa 91	U 92	Np 93	Pu 94	Am 95	Cm 96	Bk 97	Cf 98	Es 99	Fm 100	Md 101	No 102	Lr 103	Rf 104	Db 105	Sg 106	Bh 107	Hs 108	Mt 109	Ds 110	Rg 111	Cn 112	Nh 113	Fl 114	Mc 115	Lv 116	Ts 117	Og 118

图6　长式周期表

回到用同一个字母标记的音上时，它听起来和最初的乐音很像，但又不完全一样，因为它高了八度。

元素周期变化的长度，元素重复时在本质上的近似性，都使得一些化学家不再将"律"与化学周期性挂钩。化学周期性或许不像大部分物理定律那样像一条定律。不过，我们仍可以说化学周期性提供了一个化学定律的典型例子：化学定律就是近似而复杂的，但依然会大体表现出类似定律的行为。

也许现在是讨论其他几个术语问题的好时机。周期表和周期系统有什么不同？"周期系统"在二者中范围更广。它是一个更抽象的概念，说的是元素间存在的一种基本关系。如果要把周期系统表现出来，你可以采用三维排布、环形排列或者任意一种二维表。当然，"表"这个词严格暗示了二维排列。所以，尽管现在"周期表"是律、系统和表这三个术语里最知名的，但它也是限制最多的。

元素的化学反应和元素排序

我们对元素的大部分认识，都是通过它们与其他元素的反应方式和它们的成键性质得到的。常规周期表左手边的金属与通常在右手边的非金属对立互补。用现代术语来说，这是因为金属失去电子形成正离子，而非金属获得电子形成负离子。这些电荷相反的离子结合在一起，形成氯化钠或溴化钙这样的中性盐。金属和非金属还有更互补之处。金属氧化物或氢氧化物溶于水中形成碱溶液，而非金属氧化物或氢氧化物溶于水中形成酸溶液。酸和碱通过"中和"反应生成盐和水。碱和酸就像生成它们的金属和非金属，对立但互补。

20

酸和碱与周期系统的各种起源有所关联，因为它们在当量概念中非常重要（最初用来给元素排序的就是当量）。例如，任意一种金属的当量，最初是用该金属与一定量选定的标准酸溶液反应，由所需的金属量得出。之后，“当量”这个词扩大了，表示一种元素与标准分量的氧气反应所需要的量。历史上，各周期元素的顺序由当量决定，之后由原子量决定，最后变为由原子序数（稍后解释）决定。

化学家一开始是定量比较酸碱反应的消耗量。随后，这一过程拓展到酸与金属的反应。这让化学家可以根据金属的当量设定数值标度，给它们排序。而我们之前也说了，当量不过是与一定量的酸结合的金属量。

原子量与当量不同，最早是约翰·道尔顿在19世纪初得到的。他测量结合在一起的元素化合物的质量，间接得到了原子量。但这个看上去很简单的方法也有复杂之处，道尔顿必须猜测所测化合物的化学式。问题的关键是元素的价，或者说结合力。例如，一价原子与氢原子按1∶1的比例结合，像氧这样的二价原子就以2∶1的比例结合，以此类推。

之前提过的当量，有时被视为一个纯经验的概念，因为它似乎并不取决于你是否相信原子存在。引入原子量之后，许多对原子概念感到不满的化学家都试着回到更早的当量概念。他们认为当量是纯粹的经验概念，因而更可靠。但这些愿望都是幻觉，因为当量也建立在用特定表达式来表示化合物的假设之上，而表达式是理论概念。

很多年来，当量和原子量交替使用，造成了许多混乱。道尔顿假设水由一个氢原子和一个氧原子结合而成，这可以让水的原子量与当量相等，但他猜测的氧原子价终究是错的。很多人无差别地使用“当量”和“原子量”，更加剧了混乱。当量、原子量和价之间的真正关系，直到1860年德国卡尔斯鲁厄举办了第一场科学大会才得以明确。会议澄清并广泛接受了一致的原子量的概念，为数个国家的六位学者独立发现周期系统铺平了道路，他们提出的周期表形式取得了不同程度的成功。每一份表的元素都大体按原子量增大的顺序排列。 22

在之前提到的排序概念里，第三种是原子序数，也是最现代的概念。一旦理解了原子序数，它就取代原子量成了元素的排序原理。原子序数不再依赖任何方式的结合量，我们可以从元素的原子结构入手，为原子序数做出一个简单的微观解释。元素的原子序数是由其原子核中的质子数，或者说正电荷的数量决定的。因此周期表中的每个元素都比前一个元素多一个质子。在遍历周期表的过程中，原子核中的中子数往往也会增长，这就致使原子序数和原子量大体相关，但决定任意特定元素的是原子序数。换句话说，任何特定原子总具有相同的质子数，但它们包含的中子数可以不一样。这个特点造成了同位素现象，这些变体原子就叫作同位素。

周期系统的不同表示

现代周期系统按原子序数给元素排序，使它们自然而然地分族，这种方法带来了显著的成效，但它的呈现形式也并非唯一。因此我们有很多形式的周期表，有些是为不同的用途而设计的。化学家可能喜欢能突出元素反应的形式，而电子工程师可能希望关注导电方面的相似性和模式。

周期系统的呈现方式是一个有趣的话题，也特别容易吸引大众的想象。从纽兰兹、洛塔尔·迈耶尔和门捷列夫发表早期周期表以来，人们为获得“终极”周期表做出了许多尝试。据估算，在1869年门捷列夫编制出最有名的那份表之后100年里，人们已经出版了大约700种不同的周期表。它们包括各种类型，比如三维表、螺旋柱表、同心圆表、螺旋面表、折线形表、阶梯表、镜像表等。即便是今天，研究者还在不断发表文章，展现新型或改进后的周期系统。

所有这些尝试的基础都是**周期律**，它只有一种形式。多种多样的外形也改变不了周期系统的这一点。许多化学家强调，只要满足特定的基本要求，这条定律以怎样的形式表现出来无关紧要。然而，从哲学的观点看，元素最基本的表现形式，或者说周期系统的终极形式，依旧是很重要的，尤其是它关乎应该以实际的态度考虑周期律，还是应该将其视为一个习惯问题。对此，通常的回答是，它的表现形式只不过是一种习惯。这个回答似乎与现实主义者的观点相左，他们认为，每种周期表都体现了一种性质重复方式，所以其中可能存在一个真相。

周期表的变化

1945年，美国化学家格伦·西博格提议应该将从第89号元素锕开始的元素视为与稀土元素类似的系列元素，而之前研究者认为这一系列的元素应该从第92号元素，也就是铀开始（图7）。西博格的新周期表呈现了铕（63）、钆（64）分别和当时尚未发现的第95、第96号元素的相似性。根据这种相似性，西博格成功地合成、确认了后面两种新元素，随后将它们分别命名

																H	He
Li	Be											B	C	N	O	F	Ne
Na	Mg											Al	Si	P	S	Cl	Ar
K	Ca	Sc	Ti	V	Cr	Mn	Fe	Co	Ni	Cu	Zn	Ga	Ge	As	Se	Br	Kr
Rb	Sr	Y	Zr	Nb	Mo	Tc	Ru	Rh	Pd	Ag	Cd	In	Sn	Sb	Te	I	Xe
Cs	Ba	RE	Hf	Ta	W	Re	Os	Ir	Pt	Au	Hg	Tl	Pb	Bi	Po	At	Rn
Fr	Ra	AC	Th	Pa	U												

稀土元素	La	Ce	Pr	Nd	Pm	Sm	Eu	Gd	Tb	Dy	Ho	Er	Tm	Yb	Lu

																H	He
Li	Be											B	C	N	O	F	Ne
Na	Mg											Al	Si	P	S	Cl	Ar
K	Ca	Sc	Ti	V	Cr	Mn	Fe	Co	Ni	Cu	Zn	Ga	Ge	As	Se	Br	Kr
Rb	Sr	Y	Zr	Nb	Mo	Tc	Ru	Rh	Pd	Ag	Cd	In	Sn	Sb	Te	I	Xe
Cs	Ba	LA	Hf	Ta	W	Re	Os	Ir	Pt	Au	Hg	Tl	Pb	Bi	Po	At	Rn
Fr	Ra	AC															

镧系	La	Ce	Pr	Nd	Pm	Sm	Eu	Gd	Tb	Dy	Ho	Er	Tm	Yb	Lu
锕系	Ac	Th	Pa	U	Np	Pu									

图7　西博格修正前与修正后的周期表

为镅（americium）和锔（curium）。后来，研究者又合成了20多种超铀元素。

在第三、第四行过渡元素由谁开始的问题上，标准形式的周期表也经历了一些小改动。旧周期表上的这两个元素是镧（57）

和锕（89），更新近的一些实验证据和分析则让镥（71）和铹25（103）取代了它们的位置。（见第十章）有一点很有意思，就是一些更老的周期表根据宏观性质已经提前做出了这些变动。

这些都是模糊性质的例子，或许可以称为第二级性质，与第一级性质，即元素次序不可等量齐观。在经典化学用语里，第二级性质对应一族内不同元素的化学相似性。用现代术语来说，第二级性质可以用电子构型的概念解释。不论用经典化学方法，还是用以电子构型为基础的物理方法，这一类型的第二级性质都比第一级性质更不可靠，不能作为分类的依据。确定第二级性质（按这里的定义）的方式，是用化学性质还是用物理性质作为分级依据的冲突的现代例子。更重视电子构型（物理性质）还是更重视元素的化学性质，会改变周期表中每族元素的精确位置。其实，近期关于氢在周期系统中的位置的许多争论，就是围绕着这两种方法之间的相对重要性这一问题展开的。（见第十章）

近些年来，由于人工合成了元素，元素数量已经稳稳突破了100。到本书写作时，我们已经合成了第118号元素及其之前的所有元素，并整理了它们的特征。这些合成元素是典型的不稳定元素，每次只能合成很少量原子。不过，研究者发明了精巧的化学技术，来检测这些所谓的“超重”元素的化学性质，从而查26看这些大质量原子是否具有推断的化学性质。

用更哲学的话来说，合成这些原子可以让我们查看周期律是否像牛顿的万有引力定律那样，是一条普适定律，还是说，一旦原子序数大到一定程度，它就不再出现预期中应该复现的化学性质，而是会发生一些偏离。目前还没发现意料之外的事情，不过某些超重元素是否具有预期的化学性质这个问题，还远远没有答案。周期表在这一块遇到了一个很重要的困难：相对论效应逐渐显著。（见后文）这种效应会使某些原子呈现预期外的电子构型，可能带来同样是预期外的化学性质。

理解周期表

物理学的发展对我们现在理解周期系统的行为有重要影响。现代物理学中最重要的两大理论是爱因斯坦的相对论和量子力学。

其中第一个理论对我们理解周期系统只有有限的影响，但在精确计算原子和分子方面正变得越来越重要。一旦物质以接近光速的速度运动，就需要考虑相对论。内层电子，尤其是周期系统中重原子里的那些电子，轻易就可以达到这样的相对论速度。要想精确计算原子，尤其是重原子，就必须引入相对论做修正。而且，许多看上去平平无奇的元素性质，比如金的特征颜色和汞的液态性质，都可以解释为由内层高速电子的相对论效应所致。

但到目前为止，在从理论上解释周期系统方面影响力更大的重要角色，是上文中提到的现代物理中的第二个理论。量子力学实际诞生于1900年。尼尔斯·玻尔率先将它应用到原子上，他探求的观点是，周期表中任意一族元素之间的相似性可以用它们有数量相等的外层电子来解释。电子壳层有特定电子数，这个观点在本质上很像量子的概念。它假设电子只有特定的能量或者能量包，然后它们根据自己携带的这些能量或者能量包，处于包裹原子核的这个或那个壳层之中。（见第七章）

玻尔向原子中引入量子之后，很快许多人就发展了他的理论，直到旧的量子理论让位于量子力学。（见第八章）在新的景象中，人们既将电子看作粒子，也将它们看作波。更奇怪的是这个观点：电子不再沿确定的轨迹或者说轨道绕原子核运动。人们转而用弥散的电子云来描述新景象。对周期系统的最新解释就是用电子占据了多少这样的轨道来表述的。这个解释依据的是原子的电子排列，或者说“构型”，它用轨道的占据情况表示。

这里有一个有趣的问题：化学和现代原子物理，尤其是量子力学是什么关系？大部分教材强调的主流观点认为，化学“本质上”不过是物理，所有的化学现象，尤其是周期系统，都可以在量子力学的基础上推导出来。不过，这个观点尚须考虑一些问题。例如，我们必须知道，用量子力学解释周期系统还很不完美。这一点很重要，因为化学书，尤其是教科书，往往都给人一种我们现在对周期系统的解释非常完美的印象。事实绝非如此，这点我们之后再讲。

[28] 在整个现代科学里，周期系统可以列为成果最丰富、结构最统一的思想体系之一，或许可以和达尔文的自然选择进化理论相提并论。周期表在许多人的努力下，历经将近150年的发展，如今依然处于化学研究的中心。这主要是因为它有巨大的实践价值，能够让我们预测元素各种各样的理化性质和可能的价键结构。现代化学家和化学专业的学生用不着去学100多种元素的性质，就可以根据8个主族、过渡金属，以及镧系与锕系元素中的典型成员的已知性质，做出有效预测。

我们已经列出了一些专题基础，并定义了一些关键词，下面就开始讲述现代元素周期系统发展的故事吧，这要从它诞生的[29] 18、19世纪说起。

第三章

原子量、三元素组和普劳特

人们最开始对元素进行分组，根据的是元素间的化学相似性，也就是说，根据的是元素定性的方面，而不是定量的方面。例如，金属锂、钠和钾显然拥有许多相似的性质，比如它们都很软、能够浮于水面，而且会与水发生明显的反应。最后这一点与大部分金属都不同。

但现代周期表既注重元素定性的特点也兼顾其定量的特点。早在16、17世纪，化学从整体上看就已经逐渐演变为量化的领域，也就是研究参与反应的物质**有多少**，而不是研究它们**如何**反应。安托万·拉瓦锡是让化学走上这条道路的责任人之一，他是法国贵族，后来受法国大革命波及上了断头台。拉瓦锡是率先精确测量化学反应物及其产物质量的人之一。通过测量，拉瓦锡得以摆脱一个长期存在的假说：物质燃烧与一种叫作“燃素”的物质有关。

拉瓦锡发现事实正相反。燃烧任何东西，比如某种元素的物质，所得物质的质量都会增加，不会减少。他还发现，在任何

30 化学操作前后，物质的总量都相同。发现这条“物质守恒定律”之后，又发现了其他化合定律，它们都需要更深入的解释，而其中一条定律最终导向了周期表。

拉瓦锡也没有遵循古希腊那种抽象元素的观念，将其视为各种性质的载体。反之，他着重于将元素看成化合物分解的最终产物。尽管抽象元素的观念在后来又以修正后的形式重新出现了，但与古希腊的传统彻底分割是很有必要的，尤其是中世纪炼金术士间还流传着许多神秘而不科学的观念。

1792年，德国的本亚明·里希特回归元素定量方面的工作，31 发表了一份列表，记录了后来称为当量的量（图8）。这份列表列出了不同金属与定量的某种酸（如硝酸）反应所消耗的质量。至此，人们第一次用简单的量化方法，比较了不同元素之间的性质。

碱		酸	
氧化铝	525	氢氟酸	427
氧化镁	615	碳酸	577
氨	672	癸二酸	706
氧化钙	793	盐酸	712
碳酸钠	859	草酸	755
氧化锶	1,329	磷酸	979
碳酸钾	1,605	硫酸	1,000
氧化钡	2,222	琥珀酸	1,209
		硝酸	1,405
		醋酸	1,480
		柠檬酸	1,583
		酒石酸	1,694

图8　里希特的当量表，由费舍尔在1802年修订

道尔顿

1801年，英国曼彻斯特一位年轻的教师，发表了现代原子理论的开山之作。约翰·道尔顿遵循拉瓦锡和里希特的新传统，采纳了古希腊的原子观（原子是构成任何物质的最小粒子），并将之量化。他不但假设每种元素都由特定类型的原子构成，而且着手估算它们的相对原子量。

以道尔顿利用拉瓦锡用氢气和氧气生成水的实验为例。拉瓦锡的实验表明，水包含85%的氧和15%的氢。道尔顿提出，水由一个氢原子和一个氧原子结合而成，化学式为HO。那么，假定氢原子的原子量为一个单位，那么氧原子的原子量就是85/15=5.66。氧原子的原子量其实是16，这个差异是由道尔顿没有意识到的两个问题造成的。第一，他错误地假设了水是HO，而我们都知道，水的化学式应该是H_2O。第二，拉瓦锡的数据并不十分准确。

道尔顿有关原子量的概念为定比定律提供了非常合理的解释。这条定律是说，两种元素相结合，它们结合的质量比保持恒定。这条定律可以视作两种或更多具有特定原子量的原子之间的结合按比例增加后的版本。宏观上两种元素按一定比例结合，反映的是如下事实：两种特定的原子各有特定的质量，它们不断结合在一起，因此所得产物也会呈现两者的比例。

32

其他化学家（也包括道尔顿本人）还发现了另一条化学结合律，即倍比定律。当元素A与另一种元素B结合生成多种产物时，各化合物间B的结合质量有一简单比值。例如，碳和氧结合为一氧化碳和二氧化碳，二氧化碳中结合的氧的质量是一氧化碳中的两倍。这条定律同样可以很好地用道尔顿的原子理论解释，因为它表明：在一氧化碳CO中，一个碳原子与一个氧原子结合；而在二氧化碳CO_2中，是两个氧原子与一个碳原子结合。

冯·洪堡和盖-吕萨克

现在我们再看一条化学结合律，这条定律一开始不是用道尔顿的理论解释的。1809年，亚历山大·冯·洪堡和约瑟夫·路易·盖-吕萨克发现，若用氢气和氧气反应生成水蒸气，所需氢气的体积是氧气的两倍。而且，水蒸气的体积差不多与参与反应的氢气相同。

$$2\text{体积氢}+1\text{体积氧} \rightarrow 2\text{体积水蒸气}$$

他们发现，这类行为也适用于其他气体反应，因此冯·洪堡和盖-吕萨克总结道：

> 参与化学反应的气体的体积与其气体产物的体积呈小整数比。

这条新化学定律对道尔顿的新原子理论构成了巨大的挑战。根据道尔顿的理论，任何原子都是绝对不可分的，但若假设上述气体的原子不可分，则无法解释这条定律。只有氧原子可分，上述氢和氧的反应才可能在两个氢原子和一个氧原子之间发生。

意大利物理学家阿莫迪欧·阿伏伽德罗意识到，应该是两个双原子氢分子与一个双原子氧分子结合，至此谜题才有了答案。之前没有人想到这些气体会由两个原子结合在一起，构成双原子分子。因为这些分子由两个原子构成，所以可分的是分子，而不是原子。道尔顿的理论和原子不可分性得到了维护，而且假设存在双原子气体分子，它们由两个元素相同的原子构成，那么冯·洪堡和盖-吕萨克的新定律也可以解释了。

氢和氧之间的反应是这样的：两个氢双原子分子分裂为四个原子，一个氧双原子分子分裂为两个原子。上述六个原子随即形成两个水蒸气分子，即H_2O。事后看来，这些都非常简单，但考虑到双原子分子是一种激进的观点，加上当时还不知道水分子的结构式，那么也无怪这样一个简单的方程式

$$2H_2 + O_2 \rightarrow 2H_2O$$

研究者花了50年才完全理解。

但因为一段奇怪的历史波折，道尔顿本人拒不接受双原子分子的设想，因为他坚信同一元素的任意两个原子都应该相互排斥，它们绝不可能形成双原子分子。任意两个相似原子之间的化学键还是个新概念，需要人们慢慢去适应，尤其是对道尔顿这样的人来说，他们对原子应该是什么样的有着相当复杂的观点。同时，像阿伏伽德罗这样的人就能迅速前进，不被两个相似原子会互相排斥这样的想法支配（现在我们知道，原子实际上并不如此），从而提出双原子分子的假设。

安德烈·安培（电流的单位安培就是以他的名字命名的）也独立提出了阿伏伽德罗的双原子分子设想。但这个重要发现休眠了大约50年，才终于在另一位住在西西里的意大利人斯塔尼斯奥拉·坎尼扎罗的手里重见天日。

普劳特假说

道尔顿等人发表原子量数据数年后，苏格兰内科医生威廉·普劳特注意到一个有趣的现象。人们确定的很多元素的原子量，似乎是氢原子量的整数倍。他于是得出一个明显的结论：也许所有原子都只是由氢原子构成的。如果属实，那也就意味着所有物质在根本上是一致的，对这个观念，人们从古希腊哲学伊始就做过不太认真的考虑，并以不同形式重现了许多次。

但不是所有发表的原子量都正好是氢原子量的整数倍。普劳特没能解释这个特征，但他猜测这些原子量反常的原因在于没有得到精确测量。我们后来知道，普劳特的猜想带来了很丰富的成果，因为其他人因此越来越精准地测量原子量，就为了验证他是对是错。研究者相继得到越来越精确的原子量，它们最终在周期表的发明和演变中扮演了关键角色。

不过，对普劳特猜想最早的共识是它不正确。更精确的原子量测量结果表明，一般原子的原子量并不是氢原子量的整数倍。然而，很久之后，普劳特猜想还会如注定一般回归，只不过是以修正后的样子。

德贝赖纳的三元素组

德国化学家沃尔夫冈·德贝赖纳发现了另一条一般定律，这条定律也促使人们更精确地测量原子量，并由此为周期表的编制铺平道路。那得从1817年讲起，德贝赖纳发现元素中存在各种各样的组合，每个组合中的一种元素具有另两种元素的性质，而原子量大致是另两种元素的平均值。这些三种元素的组合后来被叫作三元素组。例如，锂、钠和钾都是灰色、质软、密度低的金属。锂与水的反应不剧烈，而钾和水会剧烈反应。钠则显示出居于三元素组中另外两者之间的反应强度。

此外，钠的原子量（23）也在锂（7）和钾（39）的正中间。这个发现意义重大，因为它第一次提示我们，数字规律蕴含在元素的本质与性质之关系的核心。它表明了元素间化学联系背后的数学秩序。

德贝赖纳找出的另一组关键三元素组由氯、溴和碘这三个卤族元素组成。但德贝赖纳并没有尝试将这些三元素组用某种方式联系起来。如果他这么做了，那么他或许就会比门捷列夫等人早大概50年发明周期表。

36

德贝赖纳在辨识三元素组的过程中，要求除了上面提到的数学关系外，所找的三个元素必须在化学上相似。其他按他的路子走的研究者一开始没有像他那样挑剔，其中一些人认为自己找到了很多其他的三元素组。例如，1857年，在威斯巴登工作的二十岁的德国化学家恩斯特·伦森发表了一篇文章，他把全部58种已知元素分成了20个三元素组。其中10个三元素组由非金属和成酸金属组成，剩下的10个组都是金属。

伦森借助图9的20个三元素组，又宣布辨识出总共7个超级三元素组，它们以3个三元素组为一族，每族中间的三元素组的平均当量大致等于另两组的平均当量。可以说，它们是三元素组的三元素组。但伦森的系统多少有点勉强。例如，他将单个元素氢算作一个三元素组，代替真正的三元素组，因为他觉得这么做很方便。而且，他发表的三元素组中，有很多虽然从数字上看很不错，但它们内部没有任何化学相似性。伦森和其他一些化学家被明显的数字关系诱惑，却把化学抛诸脑后了。

另一种元素分类系统来自1843年在德国工作的利奥波德·格梅林。他发现了一些新的三元素组，并着手将它们联系起来，形成一套形状独特的整体分类系统（图10）。他的系统包含55种元素，并率先按原子量递增的顺序排列了大部分元素，但他没有明确地表达出这种理念。

37

然而，格梅林的系统并不能看作周期表，因为它没有表现出元素性质的重复性。换言之，它还没有呈现周期表赖以得名的化学周期性特点。后来，格梅林用他的元素系统组织编写了一本500多页的化学教科书。

这也许是元素表第一次被用作整本化学书的基础，这在今天已是标准操作，但我们仍要注意那时用的还不是**周期**表。

克雷默斯

现代周期表不只是把元素集成一族族，以显示它们的化学相似性。除了所谓的“纵向关系”（这体现在三元素组中），现代周期表还将元素按序排列。

周期表包括竖直方向上的相似元素，也包括水平方向上的不相似元素。第一位思考水平方向关系的人是德国科隆的彼得·克雷默斯。他在一个包括氧、硫、钛、磷和硒的元素小系列里，注意到一个规律（图11）。

克雷默斯还发现了一些新的三元素组，比如：

$$Mg = \frac{O + S}{2},\ Ca = \frac{S + Ti}{2},\ Fe = \frac{Ti + P}{2}$$

	原子量计算值			原子量测量值		
1	(K + Li)/2	= Na	= 23.03	39.11	23.00	6.95
2	(Ba + Ca)/2	= Sr	= 44.29	68.59	47.63	20
3	(Mg + Cd)/2	= Zn	= 33.8	12	32.5	55.7
4	(Mn + Co)/2	= Fe	= 28.5	27.5	28	29.5
5	(La + Di)/2	= Ce	= 48.3	47.3	47	49.6
6	Yt Er Tb			32	?	?
7	Th混合稀土金属 Al			59.5	?	13.7
8	(Be + Ur)/2	= Zr	= 33.5	7	33.6	60
9	(Cr + Cu)/2	= Ni	= 29.3	26.8	29.6	31.7
10	(Ag + Hg)/2	= Pb	= 104	108	103.6	100
11	(O + C)/2	= N	= 7	8	7	6
12	(Si + Fl)/2	= Bo	= 12.2	15	11	9.5
13	(Cl + J)/2	= Br	= 40.6	17.7	40	63.5
14	(S + Te)/2	= Se	= 40.1	16	39.7	64.2
15	(P + Sb)/2	= As	= 38	16	37.5	60
16	(Ta + Ti)/2	= Sn	= 58.7	92.3	59	25
17	(W + Mo)/2	= V	= 69	92	68.5	46
18	(Pa + Rh)2	= Ru	= 52.5	53.2	52.1	51.2
19	(Os + Ir)/2	= Pt	= 98.9	99.4	99	98.5
20	(Bi + Au)/2	= Hg	= 101.2	104	100	98.4

图9 伦森的20个三元素组

O N H
F Cl Br I Li Na K
S Se Te Mg Ca Sr Ba
P As Sb Be Ce La
C B Bi Zr Th Al
Ti Ta W Sn Cd Zn
Mo V Cr U Mn Ni Fe
Bi Pb Ag Hg Cu
Os Ir Rh Pt Pd Au

图10 格梅林的元素表

	O		S		Ti		P		Se
原子量	8		16		24.12		32		39.62
差值		8		8		~8		~8	

图11　克雷默斯氧系列的原子量差

以现代的观点看，这些三元素组之内似乎并没有显著的化学关系。但这是因为现代的中长式周期表无法呈现一些元素之间的

39 第二级亲属性。例如，硫和钛虽然在中长式周期系统中不在同一族，但它们都表现为四价。不过，认为它们在化学上相似并不是勉强为之。考虑到钛和磷一般表现为三价的事实，这种分组并不像现代读者认为的那样错得离谱。但总的来说，克雷默斯和伦森的情况一样，属于不计代价地尝试创造新的三元素组。他们的目标似乎是从元素的原子量关系中找出三元素组，而不管它们在化学上是否相似。后来，门捷列夫将同行的这种做法形容为三元素组迷局，认为这延误了成熟周期系统的出现。

但回到克雷默斯，他最突出的贡献在于提出了一种双向表格，他称之为"共轭三元素组"。其中，某些元素同时是两个相交的不同三元素组的成员。

Li 6.5	Na 23	K 39.2
Mg 12	Zn 32.6	Cd 56
Ca 20	Sr 43.8	Ba 68.5

由此，克雷默斯以比前人更深入的方式比较化学上**不相似**的元素，这一研究方法直到洛塔尔·迈耶尔和门捷列夫发表周期表

40 才完全成熟。

第四章

迈向周期表

19世纪60年代是发现周期表的重要十年。它起始于在德国卡尔斯鲁厄召开的一场会议，会议的目的是解决几个有关化学家如何理解原子和分子概念的问题。

第二章中提到过，盖-吕萨克发现的气体化合体积定律，只有假设存在由两个原子结合成的可分双原子分子（比如H_2、O_2等）才能得以解释。因为道尔顿等人反对，这个假设没有被普遍接受。而在卡尔斯鲁厄会议上，由于坎尼扎罗倡导，这个观点终于被广泛接受了。坎尼扎罗是阿伏伽德罗的同乡，他在50年前就首次提出了这个观点。

另一个问题是，许多研究者给出的元素原子量都不一样。坎尼扎罗又成功计算出一套能合理解释的值，他印刷了手册，分发给参会的代表。有了这些改变，再稍待些时日，就有多达六位科学家各自提出初步的周期系统，这些系统包含了当时已知的60多种元素里的大部分。

41

德·尚古尔多阿

第一位真正发现化学周期性的人是法国地质学家亚历山大-埃米尔·贝吉耶·德·尚古尔多阿。他将元素按原子量递增的顺序排列，以螺旋状刻在一根金属柱上。他注意到化学性质相似的元素都落在一条条竖线上，与环绕圆柱的螺旋线相交（图12）。最根本的发现是，若将元素按自然顺序排列，则每隔一段就会出现相近的元素。就像一星期中的天、一年中的月、音阶中的音那样，周期性或重复性似乎是元素的本质属性。化学重复的背后成因，在之后的许多年里依旧是一个谜。

德·尚古尔多阿表示自己支持普劳特猜想，而且更甚一步，他将自己表里的原子量约成了整数。钠的原子量是23，在他的系统中距离原子量为7的锂一个循环。他把镁、钙、铁、锶、铀和钡放到下一列。在现代周期系统中，其中的镁、钙、锶和钡确实在同一族。他把铁和铀包括在内，初看上去似乎错得离谱，但我们之后会看到，很多早期的短式周期表的特点，就是将一些过渡元素与现在所说的主族元素放在一起。

这位法国人也有点走背运，他的第一篇，也是最重要的一篇论文没有收录他的系统图样。这真是一个严重的疏忽，因为在所有周期系统中，图表展示都相当重要。为了弥补这个错误，德·尚古尔多阿之后又自费发表了自己的论文，但结果也没能广泛传播，在当时的化学界始终无人知晓。德·尚古尔多阿的关键发现不为化学家所见，也因为他不是他们中的一员，而是一位地质学家。最后，这项发现没有得到关注，还因为它领先于时代。

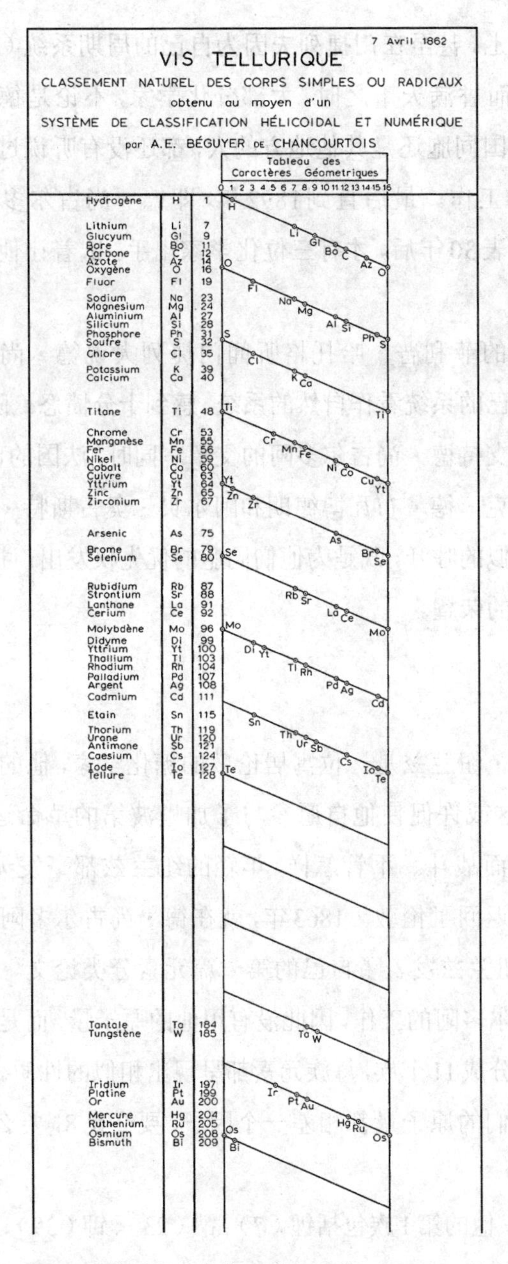

图12　德·尚古尔多阿的地螺旋

实际上，甚至在门捷列夫因为自己的周期系统（1869年开始发表）而誉满天下之时，大部分化学家，不论是德·尚古尔多阿的法国同胞还是其他地方的人，都还没有听说过德·尚古尔多阿的工作。最后直到1892年，即德·尚古尔多阿的开创性论文发表30年后，才有三位化学家出手，试着让他的研究重见天日。

英国的菲利普·哈托格听闻门捷列夫说德·尚古尔多阿没有把自己的系统看作自然的系统，感到十分愤怒。随后，他发表了一篇支持德·尚古尔多阿的文章。同时，法国的保罗-埃米尔·勒科克·德·布瓦博德朗和阿尔贝·奥古斯特·拉帕朗也发出了类似的呼吁，既是为他们同胞的优先权发出声援，也是为维护高卢的荣耀。

纽兰兹

约翰·纽兰兹是一位寓居伦敦的糖化学家，他的母亲是意大利裔，这或许促使他自愿参与了加里波第的革命运动，为统一意大利而战斗。不管怎样，年轻的纽兰兹都毫发无伤，因为他很快就返回了伦敦。1863年，就在德·尚古尔多阿发表论文一年后，纽兰兹发表了自己的第一篇元素分类论文。他不知道德·尚古尔多阿的工作，因此没有用他的原子量，而是将当时已知的元素分成11个族，每族元素都呈现出相似的性质。此外，他注意到它们的原子量都相差一个因子，要么是8，要么是8的倍数（图13）。

例如，他的第1族包括锂（7）、钠（23）、钾（39）、铷（85）、铯（123）和铊（204）。以现代的观点看，他只把铊放错了位置，第1族 碱金属：锂，7；钠，23；钾，39；铷，85；铯，123；铊，204。

本族元素当量间的关系（见《化学新闻》，1863年1月10日），这样写或许是最简单的，即：

1份锂 ＋ 1份钾 ＝ 2份钠。

1 ″ ＋ 2 ″ ＝ 1份铷。

1 ″ ＋ 3 ″ ＝ 1份铯。

1 ″ ＋ 4 ″ ＝ 163，一种尚未发现的金属的当量。

1 ″ ＋ 5 ″ ＝ 1份铊。

第2族 碱土金属：镁，12；钙，20；锶，43.8；钡，68.5。

在本族中，锶是钙和钡的平均数。

第3族 土族金属：硼，6.9；铝，13.7；锆，33.6；铈，47；镧，47；didymium[①]，48；钍，59.6。

铝等于2份硼，或硼与锆之和的三分之一。（铝也是锰的一半，锰与铁、铬构成倍半氧化物，和矾土同质同象。）

1份锆 ＋ 1份铝 ＝ 1份铈。

1 ″ ＋ 2 ″ ＝ 1份钍。

镧、didymium与钍相等或几乎相等。

第4族 氧化物与镁砂同质同象的金属：镁，12；铬，26.7；锰27.6；铁，28；钴，29.5；镍，29.5；铜，31.7；锌，32.6；镉，56。

在本族两端的镁和镉之间，锌是平均数。钴和镍相等。在钴和锌之间，铜是平均数。铁是镉的一半。在铁和铬之间，锰是平均数。

第5族 氟，19；氯，35.5；溴，80；碘，127。

在本族中，溴是氯和碘的平均数。

第6族 氧，8；硫，16；硒，39.5；碲，64.2。

在本族中，硒是硫和碲的平均数。

第7族 氮，14；磷，31；砷，75；锇，99.6；锑，120.3；铋，213。

图13　纽兰兹1863年元素分族中的前七族

① 后来证明是钕和镨的混合物。

铊应该属于硼、铝、镓、铟一族。铊元素一年前刚由英国人威廉·克鲁克斯发现。第一个正确地把铊列入硼族的人，是周期系统的共同发现者、德国的尤利乌斯·洛塔尔·迈耶尔。即使是伟大的门捷列夫在早期的周期表里也放错了铊的位置，他像纽兰兹一样，把铊放进了碱金属里。

纽兰兹在他的第一篇元素分类文章中，对碱金属做出了下面的评论：

> 本族元素当量间的关系，这样写或许是最简单的，即1份锂（7）+1份钾（39）=2份钠。

当然，这不过是重新发现了这些元素的三元素组关系，因为

Li	7	
Na	23	2Na（23）= 7 + 39
K	39	

1864年，纽兰兹开始发表一系列文章，用自己的方式摸索出一个更完善的元素系统，并发展成他之后所称的“八音律”，也就是元素每隔8个就重复一次的观点。1865年，他将65种元素收录到自己的系统里，并用原子序数作为排序的依据，而不再按原子量递增的顺序排列。现在，他开始相当自信地写下一条新的定律，而德·尚古尔多阿曾简单地考虑这条定律是否可能成立，但还是否决了它。

由于纽兰兹的“八音律”采用了音乐类比，并且他也不是科班出身的化学家，所以他在1866年向皇家化学学会口头陈述结果时（图14），给人异想天开的印象。在这群不苟言笑的听众

No.	No.	No.	No.	No.	No.	No.	No.
H 1	F 8	Cl 15	Co & Ni 22	Br & Ni 22	Pd 36	I 42	Pt & Ir 50
Li 2	Na 9	K 16	Cu 23	Rb 30	Ag 37	Cs 44	Os 51
G 3	Mg 10	Ca 17	Zn 24	Sr 31	Cd 38	Ba & V 45	Hg 52
Bo 4	Al 11	Cr 19	Y 25	Ce & La 33	U 40	Ta 46	Tl 53
C 5	Si 12	Ti 18	In 26	Zr 32	Sn 39	W 47	Pb 54
N 6	P 23	Mn 20	As 27	Di & Mo 34	Sb 41	Nb 48	Bi 55
O 7	S 14	Fe 21	Se 28	Ro & Ru 35	Te 43	Au 49	Th 56

图14　纽兰兹1866年向皇家化学学会展示八音律的图表

中，有一名会员问他，有没有考虑过按字母顺序排列元素。纽兰兹的论文没有在学会的会志上发表，不过他在其他几份化学期刊上发表了进一步的论文。虽然这种观点已经在一些化学家头脑中产生了，但接受它的时刻尚未到来。不过纽兰兹坚持下来了，不停回复反对者，还发表了不少其他的周期表。

奥德林

另一位发表过早期周期表的化学家是威廉·奥德林。他和纽兰兹不同，是位顶尖的学院派化学家。奥德林参加了卡尔斯鲁厄会议，成为德·尚古尔多阿观点在英国的拥护者。他还身兼多个要职，比如牛津大学的化学教授、伦敦阿尔伯马尔大街上皇家研究院的院长等。奥德林独立发表了自己版本的周期表，他和纽兰兹一样将元素按原子量递增顺序排列，并在横行表现元素相似性（图15）。

奥德林在1864年的一篇论文中写道：

> 将已知的60多种元素按各自不同的原子量或比重数排列，我们从得到的算术数列中可以看到显著的连续性。

接下来是进一步的评论：

> 将元素根据通常的分族做水平排列，可以轻松地得到这个纯粹的算术数列。前三列很完美，后两列有些不规则，但差别很小，微不足道。

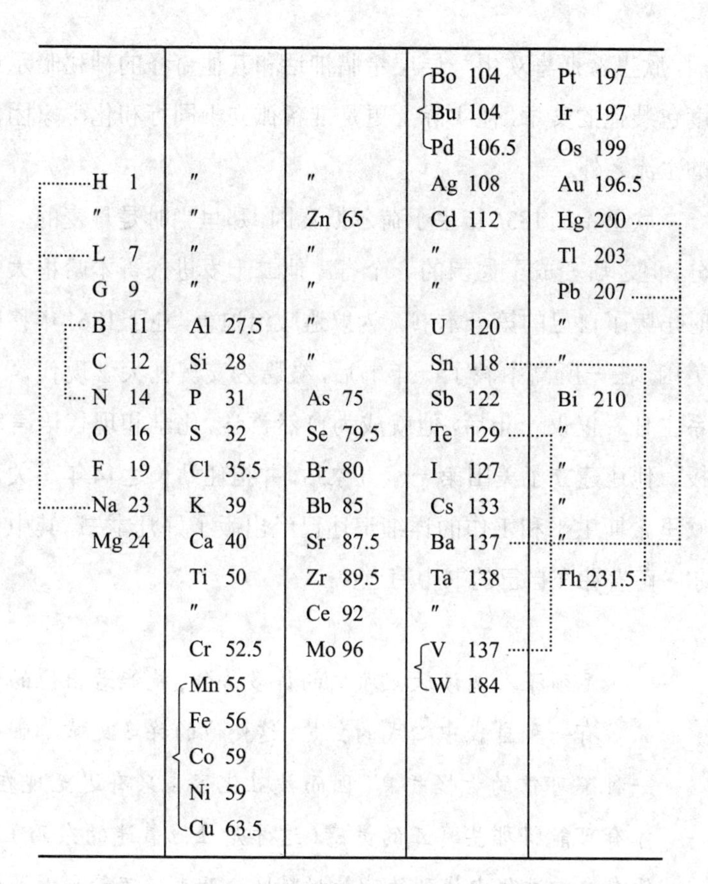

			Bo 104	Pt 197
			Bu 104	Ir 197
			Pd 106.5	Os 199
H 1	″	″	Ag 108	Au 196.5
″	″	Zn 65	Cd 112	Hg 200
L 7	″	″	″	Tl 203
G 9	″	″	″	Pb 207
B 11	Al 27.5	″	U 120	″
C 12	Si 28	″	Sn 118	″
N 14	P 31	As 75	Sb 122	Bi 210
O 16	S 32	Se 79.5	Te 129	″
F 19	Cl 35.5	Br 80	I 127	″
Na 23	K 39	Bb 85	Cs 133	″
Mg 24	Ca 40	Sr 87.5	Ba 137	″
	Ti 50	Zr 89.5	Ta 138	Th 231.5
	″	Ce 92	″	
	Cr 52.5	Mo 96	V 137	
	Mn 55		W 184	
	Fe 56			
	Co 59			
	Ni 59			
	Cu 63.5			

图15　奥德林的第三份周期表

我们不清楚为什么奥德林的发现没有被接受，因为他并不缺少学术能力的证明。看上去这或许是因为奥德林自己对化学周期性缺乏热情，不怎么相信它可能是一条自然定律。

欣里希斯

在美国，一名新来的丹麦移民古斯塔夫·欣里希斯正忙着研究自己的元素分类系统，他发表了一版相当激进的成果。然而，欣里希斯喜欢用一层层希腊神话和其他奇怪的神秘暗示过度包装他的文章，他实际上更是准备孤立于同行和化学家团体的主流之外。

欣里希斯1836年生于荷尔斯泰因，那里当时是丹麦的一部分，但之后变成了德国的一个省。他二十岁进入哥本哈根大学时出版了自己的第一本书。为躲避政治迫害，他于1861年移民美国，在一所高中教了一年书后，被聘为艾奥瓦大学现代语言系主任。仅仅一年后，他就成为自然哲学、化学和现代语言教授。他还建立了美国第一座气象站，并担任站长达14年。关于欣里希斯生平和工作的详细记述已经出版了好几本书，其中卡尔·萨普弗在自己的书中写道：

49

> 无须深入阅读欣里希斯的许多著作，就能看出他的作品多有一种自我中心式的狂热，这使他的许多文章都带着一种不可信的古怪气息，因而被丑化了。只有迟至现在，才有可能将那些真正的灵感（这才是让他着迷的东西）与他在自学过程中找到的背景材料区分开来。不管来源是什么，欣里希斯通常都会摆出好几种语言，以示炫耀。这种伪装强烈到他甚至把古希腊哲学视为自己的学说。

欣里希斯广泛的兴趣还延伸到天文学。他像许多之前的作家（可以追溯到柏拉图）一样，注意到了一些有关行星轨道大小的数字规律。他在1864年发表的一篇文章中列出了图16所示的表格，并随后做出了解释。

欣里希斯用公式$2^x \times n$来表示这些距离之差，其中n是水星

至太阳的距离	
水星	60
金星	80
地球	120
火星	200
小行星	360
木星	680
土星	1,320
天王星	2,600
海王星	5,160

图16　欣里希斯的行星距离表（1864年）

和金星距离太阳之差，即20个单位。根据x的值，套用公式可以得出如下距离：

$$2^0 \times 20 = 20$$
$$2^1 \times 20 = 40$$
$$2^2 \times 20 = 80$$
$$2^3 \times 20 = 160$$
$$2^4 \times 20 = 320$$
等等

50

在此前几年的1859年，德国人古斯塔夫·基尔霍夫和罗伯特·本生发现，可以让每种元素发光，将这些光用棱镜散射并量化分析。他们还发现，每种元素都有独特的谱图，由一系列特定的谱线构成。他们测定了谱线，并发表为一份精心制作的表格。一些人提出，这些谱线可能提供了形成它们的元素的信息，不过这些见解遭到了发现者本生的激烈批评。实际上，本生一直非常反对为了获得原子的信息或用某种方式给原子分类而研究谱线。

然而，欣里希斯毫不犹豫地将谱图与相应元素的原子联系起来。他尤其对一个现象越来越感兴趣：对每个元素来说，它的谱线频率似乎总是最小频率差的整数倍。以钙为例，它的谱线频率中就能看到1：2：4这一比例。欣里希斯对这个现象的解释既大胆又优雅：既然行星轨道的大小产生了一系列有规律的

51 整数（之前提到过），那么如果谱线的差值也是整数比，后者的原因或许就藏在不同元素原子大小的比例之中。

欣里希斯的轮形系统从中心辐射出十一根“轮辐”，包括三大非金属族和八个包含金属的族（图17）。以现代的眼光看，非金属族的顺序似乎是错的，因为在螺旋的顶端从左到右，它的顺

52 序是16，15，然后是17。包含碳和硅的那一族被欣里希斯分到了金属族里，大概是因为它还包含了金属镍、钯和铂。在现代的表里，这三种金属确实是一族的，但不和碳、硅同族。碳和硅属于锗、锡和铅所在的第14族。

不过，总的来说，欣里希斯的周期系统在许多重要元素的分类上还是相当成功的。它的一个主要优点是，比起诸如纽兰兹1864年和1865年更精细但不那么成功的周期表来说，它在分族上更明晰。欣里希斯在矿物学方面很专业，也有很深厚的化学知识。他或许是所有发现周期系统的人中跨学科幅度最大的。欣里希斯沿着一条和其他人非常不同的路子提出了他的系统，这件事或许可以为周期系统提供独立的支持。

欣里希斯在一篇1869年发表于《药剂师》上的文章中讨论了之前在元素分类方面的失败尝试。但他完全没有提到德·尚

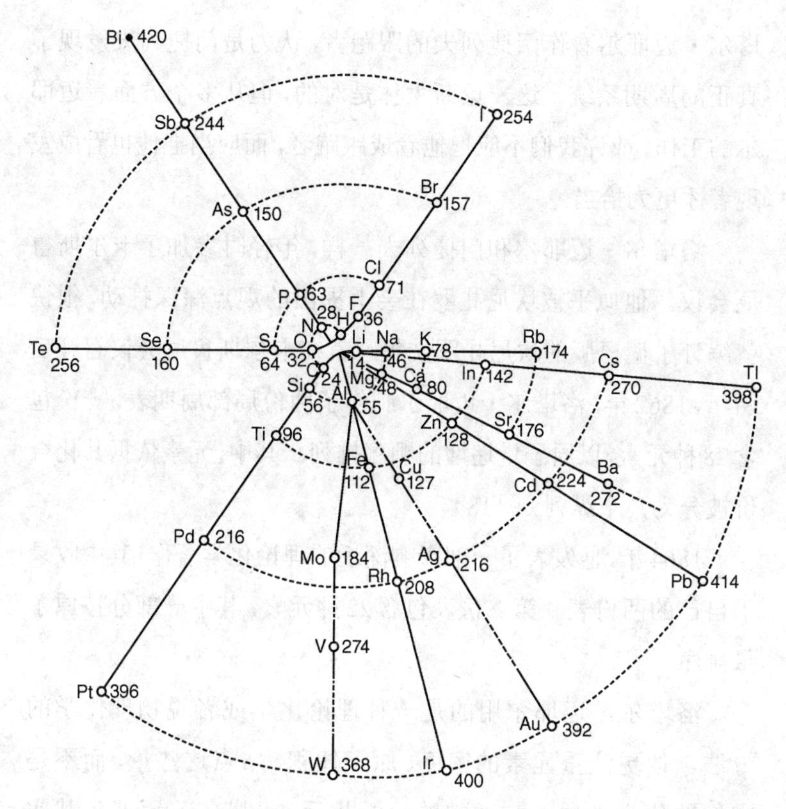

图17 欣里希斯的周期系统

古尔多阿、纽兰兹、奥德林、洛塔尔·迈耶尔和门捷列夫等发现者。欣里希斯看上去像是完全忽略了其他人直接以原子量为依据来分类元素的尝试，而有人认为，既然他懂得几门外语，那么他本应该能够注意到那些工作。

洛塔尔·迈耶尔

第一个对科学界产生了一定影响的周期系统，是德国耶拿的化学家尤利乌斯·洛塔尔·迈耶尔提出的。但人们一般将洛塔尔·迈耶尔看作门捷列夫的跟跑者，认为是门捷列夫发现了真正的周期系统。这一论断大体是对的，但从多个方面看迈耶尔的工作，或许我们不应把他看成跟跑者，而应当把他也看成发现者才更为恰当。

53

洛塔尔·迈耶尔和门捷列夫一样，年轻时参加了卡尔斯鲁厄会议。他似乎被坎尼扎罗在会上提出的观点深深打动，很快就编订了德语版的坎尼扎罗文集。就在卡尔斯鲁厄会议召开后两年，1862年，洛塔尔·迈耶尔制作了两份局部周期表，一份包含28种元素，以原子量递增的顺序排列。其中，元素依据其化合价被分为几个纵列（图18）。

1864年，他发表了一部影响深远的理论化学著作，其中收录了自己的两份表。第二份表包含22种元素，其中一部分按原子量排序。

洛塔尔·迈耶尔用的是一种理论化学或者说物理化学的方法。他更注重元素的密度、原子体积、熔点这些量，而不是它们的化学性质。与通常的论断相反，洛塔尔·迈耶尔其实在他的周期表中留了空位，甚至还试着预测了一些将会填入其中的元素的性质。其中一个预测是1886年分离出来、命名为锗的元素。洛塔尔·迈耶尔和门捷列夫不同，他相信所有物质本质上都是同一的，支持普劳特关于元素本质是复合体的假说。

1868年，他为自己那部教材的第二版制作了一个扩展了的周期系统，包含当时已知的53种元素（图19）。不幸的是，这份表被出版商弄丢了。它没有出现在新版教材中，也没有发表在任何期刊上。甚至洛塔尔·迈耶尔本人都似乎忘记了这份表，

	四价	三价	二价	一价	一价	二价
	--	--	--	--	Li = 7.03	(Be = 9.3?)
差=	--	--	--	--	16.02	(14.7)
	C = 12.0	N = 14.04	O = 16.00	Fl = 19.0	Na = 23.05	Mg = 24.0
差=	16.5	16.96	16.07	16.46	16.08	16.0
	Si = 28.5	P = 31.0	S = 32.07	Cl = 3S.46	K = 39.13	Ca = 40.0
差=	$\frac{89.1}{2} = 44.55$	44.0	46.7	44.51	46.3	47.6
	--	As = 75.0	Se = 78.8	Br = 79.97	Rh = 85.4	Sr = 87.6
差=	$\frac{89.1}{2} = 44.55$	45.6	49.5	46.8	47.6	49.5
	Sn = 117.6	Sb = 120.6	Te = 128.3	I = 126.8	Cs = 133.0	Ba = 137.1
差=	89.4 = 2 × 44.7	87.4 = 2 × 43.7	--	--	(71 = 2 × 35.5)	--
	Pb = 207.0	Bi = 208.0	--	--	(Tl = 204?)	--

图18　洛塔尔·迈耶尔1862年的周期系统

1	2	3	4	5	6	7	8
		Al = 27.3 $\frac{28\ 7}{2} = 14.8$	Al. = 27.3				C = 12.00 16.5 Si = 28.5 $\frac{89.1}{2} = 44.55$
Cr = 52.6	Mn = 55.1 49.2 Ru = 104.3 92.8 = 2.46.4 Pt = 197.1	Fe = 56.0 48.9 Rh=103.4 92.8=2.46.4 Ir = 197.1	Co = 58.7 47.8 Pd = 106.0 93 = 2.465 Os = 199.	Ni = 58.7	Cu = 63.5 44.4 Ag = 107.9 88.8 = 2.44.4 Au = 196.7	Zn = 65.0 46.9 Cd = 111.9 88.3 = 2.44.5 Hg = 200.2	$\frac{89.1}{2} = 44.55$ Su = 117.6 89.4 = 2.41.7 Pb = 207.0

9	10	11	12	13	14	15	
N = 14.4 16.96 P = 31.0 44.0 AS = 75.0 45.6 Sb = 120.6 87.4 = 2.43.7 Bi = 208.0	O = 16.00 16.07 S = 32.07 46.7 Se = 78.8 49.5 Te = 128.3	F = 19.0 16.46 Cl = 35.46 44.5 Br = 79.9 46.8 I = 126.8	Li = 7.03 16.02 Na = 23.05 16.08 K = 39.13 46.3 Rb = 85.4 47.6 Cs = 133.0 71 = 2.35.5 Te = 204.0	Be = 9.3 14.7 Mg = 24.0 16.0 Ca = 40.0 47.6 Sr = 87.6 49.5 Ba = 137.1	Ti = 48 42.0 Zr = 90.0 47.6 Ta = 137.6	Mo. = 92.0 45.0 Vd = 137.0 47.0 W = 184.0	

图19　洛塔尔·迈耶尔1868年的周期系统

因为他在后来与门捷列夫爆发了居前性争论之后，都没有提过这份表。如果这份表在当时公开，那么门捷列夫的居前性声明还会不会像今天看到的那样有分量，就不好说了。

54

洛塔尔·迈耶尔遗失的那份表有突出的优点，它包含了许多元素；某些在门捷列夫同年那份著名的表上放错了位置的元素，在它上面的位置也是对的。这份遗失的表在洛塔尔·迈耶尔去世后的1895年终于发表了，但已为时过晚，它对是谁率先提出第一个完善的周期系统这个问题造不成任何冲击。

门捷列夫在公开的争议中态度更强硬，宣称荣誉只属于他一个人，因为他不只发现了周期系统，还成功做出了许多预言。洛塔尔·迈耶尔似乎采取了一种失败者的姿态，甚至承认自己没有做出预言的勇气。

57

第五章

俄国天才：门捷列夫

德米特里·伊万诺维奇·门捷列夫是迄今为止最有名的俄国近代科学家。这不仅是因为他发现了周期系统，还是因为他看出了它所指向的深层自然律——周期律。他用许多年勾画出这条定律的完整结论，尤其重要的是，预测了很多新元素和它们的性质。此外，他修正了一些已知元素的原子量，并更正了另一些元素在周期表中的位置。

但最重要的一点或许是，尽管门捷列夫在一生的多个阶段研究过其他几个非常不同的领域，但他一直在研究和改进周期表，将周期表变成了自己的成果。反观他的大部分先驱或共同发现者，他们都没有坚持自己最初的发现。于是，门捷列夫的名字与周期表牢牢地联系在了一起，就像自然选择和相对论分别与达尔文和爱因斯坦相联系一样。鉴于门捷列夫在周期表历史上特别重要，我用这一整章来回顾他的科学工作与早期发展。

门捷列夫的父亲拥有一家玻璃工厂。他在门捷列夫很小的时候就失明了，很快便去世了。德米特里是十四个孩子中的老么，由疼爱他的母亲抚养长大。母亲决心让他尽可能接受最好的教育。抱着这份决心，她带着男孩辗转几百英里，希望他能进入莫斯科大学。但他失败了，他遭拒明显是由于他出生于西伯利亚。门捷列夫的母亲没有退缩，后来成功把小门捷列夫送进了圣彼得堡的高等师范学院。他在那里学习化学、物理、生物，当然还有教育学。最后一门，即教育学对他发现成熟的周期系统有显著的作用。不幸的是，门捷列夫入学不久，母亲就去世了，留下他自谋生计。

门捷列夫完成本科学业后，先在法国待了一段时间，然后才去德国。虽然他更想待在家里自己做气体实验，但还是在德国正式加入了罗伯特·本生的实验室。门捷列夫正是在留德这段时间参加了1860年的卡尔斯鲁厄会议。这不是因为他是知名化学家，而是因为他正好在正确的时间处在正确的地点。尽管门捷列夫像洛塔尔·迈耶尔在这次会议上那样，也很快明白了坎尼扎罗观点的价值，但比起洛塔尔·迈耶尔，门捷列夫似乎花费了相当长的时间才转而使用坎尼扎罗的原子量。

1861年，门捷列夫出版了一本有机化学教材，这为他赢得了俄国的杰米多夫大奖，他自此开始展现出真正的前途。1865年，他完成了关于乙醇和水反应的博士论文答辩，着手为提高化学教学能力而写作一本无机化学教材。在这本新书的第一部分，他没有按特定的顺序排列一般元素。1868年底，他完成了第一部分，开始思考怎样过渡到第二部分剩余的元素。

真正的发现

门捷列夫在大约十年间一直思考着元素、原子量和元素分类，他似乎在1869年2月17日遇到了顿悟的一刻，或者说“顿悟的一天”。那天，他取消了以顾问身份参观奶酪厂的行程，决定思考后来成为他最有名的思想之子的周期表。

首先，他在奶酪厂的请柬背后，将几个元素符号列为两行。

Na	K	Rb	Cs
Be	Mg	Zn	Cd

然后他再列出一个稍大的，有16种元素的阵列。

F	Cl	Br	I				
Na	K	Rb	Cs			Cu	Ag
Mg	Ca	Sr	Ba	Zn	Cd		

门捷列夫一直忙到晚上，将阵列拓展到整个周期表，包含了当时已知的63种元素。而且，表上还有几个留给未知元素的空位，他甚至推测了其中一些元素的原子量。他把这第一份表印了200张，发给全欧洲的化学家。同年3月6日，门捷列夫的一位同事在俄国化学学会的会议上宣布了这个发现。一个月内，这个新成立的学会便在期刊上刊登了一篇文章，还有另一篇更长的文章在德国发表。

60

许多关于门捷列夫的通俗读物和纪录片都说，他是在梦中或是在摆弄元素的卡牌，好似在玩一种考验耐心的游戏之时发现了周期表。现在我们认为，这些说法（尤其是第二个）是门捷列夫的传记作者（如科学史学家迈克尔·戈丁）杜撰出来的。

还是回到门捷列夫的科学方法吧。他似乎与对手洛塔尔·迈耶尔非常不一样。他不相信所有物质是同一的，因此不支持普劳特关于所有元素构成本质的假说。门捷列夫也注意远离三元素组的观点。例如，他猜想元素氟应该与氯、溴、碘一起，尽管它们超过了三元素组的构成，而是至少四个元素构成一族。

一方面，洛塔尔·迈耶尔关注物理原理，而且主要关注元素的物理性质，而门捷列夫则非常熟悉它们的化学性质。另一方面，在确立元素分类应依据的最重要标准时，门捷列夫坚持以原子量为顺序，不允许有任何例外。当然，门捷列夫的许多前辈，比如德·尚古尔多阿、纽兰兹、奥德林和洛塔尔·迈耶尔，都在不同程度上认识到了原子量顺序的重要性。但门捷列夫对原子量和元素本质的哲学理解更深刻，并因此研究起了未知元素这一新的领域。

元素的本质

化学中有一个长期存在的谜题。钠和氯结合，会产生全新的物质氯化钠。参与合成的两种元素仿佛都不存在了，至少视觉上是这样的。这就是化学键，或者说化合现象，它与搅拌硫粉和铁屑这种物理混合完全不一样。

问题之一是，在化学合成中，参与合成的元素如何保留在化合物中（如果保留下来了的话）？在某些语言中（比如英语），这个问题尤其复杂，因为人们会用“元素”这个词指称化合后的物质，比如氯化钠中的氯。这样，未化合的绿色气体氯和化合后的氯有时都被称为“元素”。现在，我们对于描述周期表要归类的物质所用的这个重要化学术语，就有了三种概念。

如上所述，“元素”的第三种概念有“形而上学元素”“抽象元素”“超越元素”等不同的叫法，最近的叫法是“基本物质元素”。这就是承载了抽象性质的元素，没有诸如氯的绿色这样的可感性质。同时，“绿氯”被称为“单质”元素。

安托万·拉瓦锡在18世纪末改革了化学，他最大的贡献是关注作为单质的元素，也就是分离形态的元素。这是打算甩掉过多的形而上学包袱以改进化学，也确实取得了巨大的飞跃。这时，“元素”主要是指分离任意化合物中的成分的最后一步产物。拉瓦锡是否打算舍弃“元素”抽象和哲学的意味，是一个有争议的问题。但对能够分离出来的元素来说，这类意味无疑渐渐退居次席。

然而，研究者并没有完全忘记抽象的意味，门捷列夫就是一位不仅理解它，还提出要提升其地位的化学家。实际上，他一再强调，周期系统主要分类抽象意义的“元素”，而未必是可以分离成更具实体形态的元素。

我之所以要仔细展开探讨这个问题，是因为门捷列夫有了这一观念之后，相比那些局限于分离形态的元素的化学家，他能够以更深入的视角看待元素。这给了门捷列夫看穿表象的可能。如果某个元素看起来不适合特定的族，门捷列夫就可以利用“元素”更深层的意义；因此在某种程度上，他能够忽略元素在分离形态或单质状态下的明显性质。

门捷列夫的预测

门捷列夫最大的，或许也是最令人难忘的成就，就是正确地预测了几种新元素。此外，他还修正了一些元素的原子量，把其他一些元素换到了周期表中别的位置上。我在上一节中提到过，他有这种远见，或许是因为与竞争者相比，他对元素的本质有更深的哲学上的理解。门捷列夫将元素视为基本物质，关注其抽象意义，由此得以越过由分离的元素表面上的性质造成的明显障碍。

尽管门捷列夫认为元素的原子量最重要，但他也会考虑元素的化学、物理性质和族内元素的相似性。洛塔尔·迈耶尔专注于物理性质，而门捷列夫更关注元素的化学性质。门捷列夫采用的另一条标准是每个元素在周期表上只能占一个位置，尽管他在处理他所说的第8族时（图5）有意打破这个观点。他采用的一条更重要的标准是，根据原子量递增的顺序排列元素。然而在一两种情况下，他看似连这条原则也违背了，但深究起来，就知道实情并非如此。

碲和碘是周期表中仅有的四组调换对之一，也是其中最有名的。这对元素的位置与它们根据原子量排列本应出现的顺序相反（碲的原子量是127.6，碘是126.9，但碲排在碘之前）。许多历史概述都会做一番评论，说门捷列夫如何有智慧，在排序时将化学性质看得比原子量更重要，从而调换了元素的位置。这类评论有几方面的错误。首先，门捷列夫绝非第一位做出这一调换的化学家。在奥德林、纽兰兹和洛塔尔·迈耶尔的表里，碲和碘的位置就已经调换了，它们都比门捷列夫的文章发表得更早。其次，门捷列夫实际上并没有把化学性质看得比原子量排序更重要。

门捷列夫坚持自己按原子量递增排序的标准，并多次声明这条原则没有例外。他调换碲和碘的位置，是因为认为它们当中一个或两个的原子量测量结果并不准确，之后的工作会证明，即便按原子量顺序排列，碲也应该排在碘前面。在这件事，以及许多没有记录下来的事上，门捷列夫想错了。

在门捷列夫提出他的第一个周期系统之时，人们认为碲64和碘的原子量分别是128和127。门捷列夫相信原子量是基本排序原理，这意味着他别无选择，只能质疑这两个值的准确性。因为从化学相似性考虑，很明显碲应该和第7族的元素同族，碘应该和第8族的元素同族，换句话说，这一对元素应当“调换位置”。门捷列夫直到去世前都在质疑这两个原子量是否可靠。

一开始，他认为碘的原子量基本上是正确的，怀疑碲的原子量有问题。于是，门捷列夫在后来的一些周期表中将碲的原子量列为125。有一次，他断言最常见的128这个结果，是测量了碲与一种他称为类碲的新元素的混合物造成的。这些断言促使捷克化学家博胡斯塔夫·布劳纳在19世纪80年代初展开一系列实验，意在重新测量碲的原子量。1883年，他宣布碲的原子量应该是125。有人参加了布劳纳做出此宣告的会议，给门捷列夫发去了祝贺电报。1889年，布劳纳得到新的结果，似乎进一步支持了之前Te=125的结论。

但在1895年，整个事情又变了。布劳纳本人宣布，碲的新测原子量比碘的更大，因此事情又回到了起点。门捷列夫这时的反应是不再怀疑碲，而开始怀疑碘的公认原子量了。此时，他要求重新测定碘的原子量，希望它的值能高一点。门捷列夫在之后的一些周期表里把碲和碘的原子量都列为127。这个问题直到1913年和1914年才由亨利·莫塞莱解决，他证明元素应该按原子序数，而不是原子量排列。尽管碲比碘原子量更大，但碲的65原子序数更小，这就是碲应当放在碘之前的原因，也与其化学行为完全相符。

虽然门捷列夫的预测看上去或许不可思议，但他所依据的方法，其实是对夹在已知元素之间的未知元素的性质做仔细的插值预测。他在1871年发表的长篇论文中详细介绍了自己的预测，不过在1869年发表的第一份周期表中他就已经开始做预测了。他一开始关注的是周期表中的两个空位，一个在铝之下，一个在硅之下，他用梵语中意思为"一"或"类一"的前缀eka，分别称它们为"eka-aluminium"和"eka-boron"，即类铝和类硼。他在1869年的论文中写道：

> 我们将发现未知元素，也就是类似铝和硅的元素，其原子量在65与75之间。

1870年秋天，他开始推测第三种元素，在表中位于硼之下。他列出这三种元素的原子体积如下：

类硼	类铝	类硅
15	11.5	13

1871年，他预测它们的原子量为：

类硼	类铝	类硅
44	68	72

并且，他对这三种元素的各种化学和物理性质也给出了一套详细的预测。

六年后，这三种预测元素中的一种被首先分离出来，命名为

66 镓。除了几处细微的不同，门捷列夫的预测几乎完全正确。他预测之准也可以从被他称为类硅、后来被命名为锗的元素看出，这种元素由德国化学家克莱门斯·温克勒分离出来。门捷列夫的唯一失败之处是四氯化锗的沸点（图20）。

性　质	1871年类硅的预测值	1886年发现的锗的实际值
相对原子量	72	72.32
比重	5.5	5.47
比热	0.073	0.076
原子体积	13 cm^3	13.22 cm^3
颜色	深灰	灰白
二氧化物比重	4.7	4.703
四氯化物沸点	100℃	86℃
四氯化物比重	1.9	1.887
四乙基衍生物沸点	160℃	160℃

图20　类硅（锗）的预测性质和实测性质

门捷列夫的失败预测

门捷列夫的预测并未全部大获成功，这一点似乎被大部分记录周期表历史的通俗作品忽略了。如图21所示，在他发表的18个预测中，有8个是失败的——尽管这些预测的重要性或许并不相同。这是因为其中一些预测涉及稀土元素，而稀土元素

67 之间非常相似，之后许多年对周期表来说都是很大的挑战。

门捷列夫给出的元素	原子量预测值	原子量测量值	最终命名
氪	0.4	未发现	未发现
以太	0.17	未发现	未发现
类硼	44	44.6	钪
类铈	54	未发现	未发现
类铝	68	69.2	镓
类硅	72	72.0	锗
类锰	100	99	锝（1939）
类钼	140	未发现	未发现
类铌	146	未发现	未发现
类镉	155	未发现	未发现
类碘	170	未发现	未发现
类铯	175	未发现	未发现
三锰	190	186	铼（1925）
二碲	212	210	钋（1898）
二铯	220	223	钫（1939）
类钽	235	231	镤（1917）

图21 门捷列夫成功与失败的预测

此外，门捷列夫失败的预测还引出了另一个哲学问题。长久以来，历史学家和科学哲学家都在争论，我们应该更看重科学进步中的成功预测，还是应该更看重成功验证已知的数据。当然，有一件事没有争议，那就是成功预测会带来更大的心理冲击，因为它们相当于说该课题的研究者能够预测未来。但成功验证或解释已知的数据也是了不起的成就，尤其是我们通常有更多的已知信息可供整合进一门新的科学理论。门捷列夫和周 68

期表尤其符合这种情况，因为他必须用一套完全统一连贯的系统成功安置63种已知元素。

在周期表发现之时，诺贝尔奖还没有设立。化学界最高奖项之一是戴维奖，以化学家汉弗莱·戴维命名，由英国皇家化学学会颁发。1882年的戴维奖颁给了洛塔尔·迈耶尔和门捷列夫。这似乎表明决定奖项归属的化学家还没有完全服膺门捷列夫的成功预测，因为他们还愿意承认洛塔尔·迈耶尔，虽然迈耶尔没有做过任何预测。而且，细看他们给洛塔尔·迈耶尔和门捷列夫的颁奖词，里面也完全没有提到门捷列夫的成功预测。这也许说明，至少这一群杰出的英国化学家没有一边倒，让成功预测的心理影响压过成功排布已知元素的个人能力。

惰性气体

19世纪末发现惰性气体这件事，给周期系统带来了有趣的挑战，原因有好几个。首先，虽然门捷列夫戏剧性地预言了许多元素，但他完全没有预测到这整族元素（He、Ne、Ar、Kr、Xe、Rn）。

它们中第一个被分离出来的是氩，1894年由伦敦大学学院发现。氩与之前讨论的许多元素不同，诸多因素交织在一起，使得给这种元素排位变成了某种赫拉克勒斯的任务。直到六年后，惰性气体才取得第八个族的地位，位于卤素和碱金属之间。

69

但让我们回到第一个被分离出来的惰性气体氩。它是由瑞利勋爵和威廉·拉姆齐在研究氮气的过程中发现的，他们分离出来的量很少。要把氩放到元素周期表里，就需要其至关重要的原子量，但这不容易做到。这是因为氩分子的原子个数不容易测定。大部分测量都倾向于认为它是单原子，但当时已知的其他气体都是双原子的（H_2、N_2、O_2、F_2、Cl_2）。如果氩确实是单原子的，那么它的原子量大概就是40，于是它在元素周期表中的位置就成了问题，因为这个原子量没有余位了。元素钙的原子量大约为40，它之后的钪是门捷列夫成功预测的一个元素，原子量是44。这似乎没有给原子量为40的新元素留下空间（图5）。

氯（35.5）和钾（39）之间有一段相当大的空白，但如果把氩放在这两个元素之间就会导致刺眼的元素调换对。试回想，此时只有一对这样的重要调换，就是元素碲和碘，而这一操作可说是非常反常的。门捷列夫的结论是碲和碘调换是因为测定的碲或碘的原子量不对，或两者都不对。

氩元素另一个不寻常的地方是，它在化学性质上完全是惰性的，就是说我们无法研究它的化合物，因为它不会形成任何化合物。一些研究者认为，这种气体惰性表明它不是真实的化学元素。如果确实如此，那应该把它放在哪里的窘境就很容易解决了，因为根本没必要把它放进去。

但许多人坚持尝试把它放进周期表。如何安置氩是1885年一次皇家学会例会的重点，当时顶尖的化学家和物理学家参加了会议。氩的发现者瑞利和拉姆齐主张这种元素可能是单原子的，但也承认他们并不确定。他们也不确定这种气体是不是混合物，这或许在暗示其原子量实际上不是40。威廉·克鲁克斯提供了一些证据，支持氩具有明确的沸点和熔点，因此表明氩是单质而非混合物。顶尖化学家亨利·阿姆斯特朗主张氩或许像氮，因为尽管单个氮原子具有很高的活性，但它能形成惰性的双原子分子。

物理学家阿图尔·威廉·吕克尔主张，原子量约为40这点应该是对的，还主张如果这种元素不能纳入周期表，就是周期表本身不对。这个评论很有意思，因为它表明，即使门捷列夫的周期表已发表16年，并且他的三项著名预测都已证明为真，也不是每个人都相信周期系统是正确的。

因此，皇家学会会议对新元素氩的命运，以及它是否应该算作新元素，大体上并没有做出决定。门捷列夫本人没有参加会议，但他在伦敦的《自然》期刊上发表了一篇文章。在文中他得出结论说，氩实际上是三原子分子，由三个氮原子组成。他用假定的原子量40除以3得到13.3，与氮的原子量14相差不多，从而得到这个结论。此外，氩是在氮气实验中得到的，这使得三原子的设想貌似更合理。[71]

这个问题最终在1900年解决了。在柏林的一次会议上，新气体的发现者之一拉姆齐通知门捷列夫，当时新增了氦、氖、氪和氙四种元素的新族，可以很好地放在卤素和碱金属之间的第八列。氩是这些新元素中第一个被发现的，它的情况尤其麻烦，因为它意味着一个新的调换对。它的原子量大约为40，却出现在原子量为39的钾之前。门捷列夫这次欣然接受了提议，他后来写道：

> 确认新发现元素的位置对他［拉姆齐］非常重要，对我来说，则是周期律普适性的一次光荣验证。

惰性气体的发现非但没有威胁到周期表，其成功纳入周期表反而凸显了门捷列夫周期系统的强大力量和通用性。[72]

第六章

物理侵入周期表

尽管约翰·道尔顿重新将原子概念引入了科学，但随之也引来了化学家的争论。许多人完全不接受原子存在，门捷列夫也是持怀疑态度的化学家之一。不过，我们从前几章可以看到，这并不妨碍他提出并发表当时最成功的周期系统。进入20世纪，经过爱因斯坦等物理学家的研究之后，原子真实存在的观念已经越发坚定地树立起来。爱因斯坦1905年关于布朗运动的论文，利用统计学方法给出了原子存在的决定性理论证据，但缺乏实验支持。这很快由法国实验物理学家让·皮兰提供了。

变化由多项研究与进展共同带来，这些研究意在探索原子结构，这些进展对从理论上理解周期系统有重大影响。我们将在本章回顾这些原子研究中的一部分，以及20世纪物理学的一些其他关键发现。我把它们的贡献称作物理学对周期表的侵入。

第一种亚原子——电子——的发现和原子具有结构的最初线索都出现于1897年，出自剑桥卡文迪许实验室的传奇人物约翰·约瑟夫·汤姆生之手。稍早一些，1895年，德国维尔茨堡

73

的威廉·康拉德·伦琴发现了X射线。这种新射线很快由年轻的物理学家亨利·莫塞莱投入应用。莫塞莱起初在曼彻斯特工作,其短暂科学生涯的最后时光则在牛津度过。

伦琴发现X射线后仅一年,巴黎的亨利·贝克勒尔就发现了一个非常重要的现象。这个现象叫作放射性,具有放射性的特定原子会在分裂时释放出大量不同种类的射线。"放射性"一词实际上是由波兰裔科学家玛丽·斯克洛多夫斯卡(后来的居里夫人,图22)创造的。她与丈夫皮埃尔·居里对这种危险的新现象展开研究,很快就发现了两种新元素,他们分别称之为钋和镭。

图22 玛丽·居里

我们通过研究原子如何在分裂时发生放射性衰变，可以更有效地探测原子的组成，探讨支配原子转变为其他原子的定律。因此，尽管周期表处理的是单独的元素或其原子，但它似乎也有某些特点，能够让一些原子在适当的条件下转变为其他原子。例如，失去一个α粒子（由包含两个质子和两个中子的氦核构成），就会得到原子序数低两个单位的元素。

新西兰人欧内斯特·卢瑟福是这一时期另一位非常有影响力的物理学家。他先是去剑桥做研究员，后来又在麦吉尔大学和曼彻斯特大学待了一段时间。之后，他返回剑桥，以J. J.汤姆生继任者的身份担任卡文迪许实验室的主任。卢瑟福对原子物理的贡献是多种多样的，其中就有支配放射性衰变的定律。他还是第一个“分裂原子”的人，也是第一个将一种元素“嬗变”为新元素的人。卢瑟福由此人工实现了放射过程，以相似的方法产生了完全不同的元素原子，这再一次证明了所有物质本质上都是同一的，这刚好是门捷列夫一生全力反对的另一个观念。 74

卢瑟福还有一个发现是原子的核式模型，这个观念现在多多少少已经得到认可，它说的是，原子由中心的核与绕核运动的带负电荷电子构成。他在剑桥的前辈汤姆生，曾认为原子由构成球形的一层正电荷及其内部的电子构成，这些电子按环状运动。 75

然而，第一个提出核式模型（原子像一个小太阳系）的人并不是卢瑟福。这一荣誉属于法国物理学家让·皮兰。他在1900年提出负电子环绕正核运动，就像行星环绕太阳一样。1903年，日本的长冈半太郎给这个天文学类比带来了新变化。他提出了自己的土星模型，其中电子就像环绕土星的著名光环。但不论皮兰还是长冈，都找不到确切的实验证据来支持他们的原子模型，而卢瑟福则能够做到。

卢瑟福与后辈同事盖格、马斯登一起用一束α粒子轰击一片金箔，得到了相当出人意料的结果。虽然大部分α粒子都比较顺畅地穿过了金箔，但他们也探测到相当数量的粒子以非常大的角度被反射回来。卢瑟福的结论是，金原子或其他任何物质的原子，除一个位于中心的致密的核外，大部分都由空荡的空间构成。一些α粒子出人意料地被反弹，射向入射α粒子束，这个事实就是每个原子都存在一个致密中心核的证据。

因此，自然比人们先前认为的更灵活多变。例如，门捷列夫曾认为元素都是互不相干的。他不接受一种元素能够转变为其他元素的观点。实际上，在居里夫人逐步发表可以推测原子分裂的实验结果之后，门捷列夫曾前往巴黎，亲自查看证据，这是他去世前不久的事了。我们不清楚他在访问居里夫人的实验室之后是否接受了放射这个新观点。

X射线

1895年，五十岁的德国物理学家伦琴做出了一项重大发现。在此之前，他的研究结果还多少有些稀松平常。很久之后，原子物理学家埃米利奥·塞格雷写道："截至1895年初，伦琴已写下了48篇现在实际上已被遗忘的文章。而他的第49篇触到了金子。"

伦琴当时是在探索电流在真空玻璃管（称为克鲁克斯管）中的行为。他注意到，实验室另一头一件不在实验系统之内的物体在发光。那是一块涂有氰亚铂酸钡的屏幕。他迅速确定，发光不是由电流，而是由克鲁克斯管发出的某种新型射线引起的。很快，伦琴又发现了X射线最为人知的性质，那就是可以用来生成手的影像，清楚地显示骨头的外形。一项带来无数医学应用的新技术初见曙光。秘密研究了几星期后，伦琴准备好在维尔茨堡物理医学学会宣布结果。这是个很有意思的巧合，他的新射线给这两个当时还相距甚远的领域都带来了巨大的冲击。

伦琴的一些原始的X射线胶片被寄往巴黎，送到亨利·贝克勒尔手中。他对验证X射线和荧光性质，即一些物质暴露于阳光后会发光的性质，两者之间的关系很感兴趣。为了测试这个想法，贝克勒尔用厚纸包住一些铀盐的晶体，为避免阳光照射，他决定把这些材料在抽屉里放几天。好运再次降临，贝克勒尔偶然地把这包晶体放在了没有曝光的底片上，然后关上抽屉，在巴黎接下来的几个阴天里做自己的事去了。

打开抽屉时，他惊讶地发现，虽然没有阳光照射，铀盐却在底片上显像了。这显然说明，不考虑荧光现象，铀盐本身发出了射线。贝克勒尔发现的正是放射性，是一些材料的自然过程，其原子核自发衰变，产生能量巨大、在某些情况下非常危险的射线。几年之后，玛丽·居里将这种现象命名为“放射性”。

事实证明，这些实验与X射线存在联系的假设是错的。贝克勒尔没有找到X射线和荧光现象之间的任何联系。实际上，这些实验中完全没有X射线，但他发现的现象在不只一个方面具有重大意义。其一，放射性是探索物质和放射的最早、最重要的一步；其二，它间接地导向了核武器的研发。

回到卢瑟福

1911年前后，卢瑟福在分析了原子散射实验的结果后，得到原子核电荷大约是原子量一半的结论，即$Z \approx A/2$。牛津物理学家查尔斯·巴克拉也支持这条结论，他以一种完全不同的方法，用X射线的相关实验得到了同样的结论。

同时，对这个领域完全外行的荷兰计量经济学家安东·范登布鲁克，思考着是否有可能修正门捷列夫的周期表。1907年，他提出了一份包含120种元素的新表，其中留有许多空位。一些空位被新发现的物质占据，它们是不是元素仍存有疑问，比如所谓的钍放射物、铀X（铀的一种未知衰变产物）、Gd_2（钆的一种衰变产物）等许多新种类。

但范登布鲁克的工作的真正创新点，是提出了所有元素都由一种他称为"阿尔法子"的粒子组成。阿尔法子由具有两个原子量单位的半个氦原子构成。1911年，他发表文章做进一步阐述，他去掉了所有对阿尔法子的表述，但保留了元素相差两个原子量单位的观点。他在写给伦敦《自然》期刊的一封二十行短信中，以下面这句话进一步探讨了原子序数的概念："潜在元素的序数等于潜在永久电荷的序数。"

范登布鲁克因此提出，既然原子的核电荷是原子量的一半，而连续元素的原子量以两个单位阶梯式递增，那么核电荷就决定了元素在周期表中的位置。换言之，周期表中每个连续元素都应该比它之前的元素多一个核电荷。

1913年，范登布鲁克又发表了一篇论文，它吸引了尼尔斯·玻尔的注意，玻尔将这篇论文列为1913年度他最喜欢的有关氢原子和多电子原子的电子构型的三篇论文之一。同年，范登布鲁克的另一篇文章也在《自然》上发表，文中他明确将每个原子的序数与该原子的电荷联系起来。也许更重要的是，他没有把序数与原子量相联系。这篇里程碑式的文章得到了领域内许多专家的赞扬，包括索迪和卢瑟福，他们都没有像业余的范登布鲁克这样，分析得这么清楚。

莫塞莱

尽管业余选手在原子序数的概念上胜出专家一筹，但他并没有完成确定这个新量的任务。完成了这项任务，并且获得发现原子序数之人美誉的是二十六岁时死于第一次世界大战的英国物理学家亨利·莫塞莱。他仅仅靠一篇两页长的文章就获得了声名。文中，他用实验证实，对元素而言，原子序数是比原子量更好的排序原则。这项研究的重要之处还在于，让我们能够知道自然存在的元素（位于氢和铀之间）中还有多少有待发现。

莫塞莱在曼彻斯特大学接受学业训练，是卢瑟福的学生。莫塞莱的一项实验是让光在不同元素的样本表面反射，记录每种样本发出的X射线特征频率。样本会发出X射线，是因为内层电子从原子射出，导致外层电子填入空位，这个过程就伴随着X射线的发出。

莫塞莱先选择了14种元素，其中9种是钛到锌，它们在周期表上构成一段连续的元素序列。他发现，发出的X射线的频率与代表每个元素在周期表中位置的整数的平方，在坐标系上构成一条直线。这就证实了范登布鲁克的假说，即元素可以用之后被称为原子序数的整数的序列来排序。元素从H=1、He=2开始，一直排列下去。

80

在第二篇论文中，他将这个关系扩展至另外30种元素，进一步巩固了其地位。至此，判断许多新公布的元素是否有效，对莫塞莱来说就很简单了。例如，日本化学家小川诚二曾宣布他分离出了一种元素，能够填入周期表中锰下方的空位。莫塞莱用电子轰击小川的样本，测量了X射线的频率，发现它的值并不符合对第43号元素的预期。

在化学家用原子量来排列元素时，有多少种元素还有待发现是非常不确定的。这是因为，周期表中连续元素的原子量的差值很不规律。换成原子序数后，这种复杂性就消失了。现在，连续元素的差值相当规律，就是一个单位的原子序数。

莫塞莱去世后，其他化学家和物理学家采用他的方法，根据均匀分布的原子序数，发现了当时未知的元素，它们的原子序数分别为43、61、72、75、85、87和91。直到1945年研究者合成了第61号元素钷，最后一个空缺才填上。

同位素

发现一种特定元素的同位素是理解周期表的另一个关键步骤。它发生在原子物理的初始阶段。同位素（isotope）这个词来源于iso（相同的）与topos（位置），用于描述任意一种元素的原子种类，它们的原子量不同，但在周期表中占据同一位置。这个发现在一定程度上是必然的。原子物理的新发展让研究者发现了大量新元素，如Ra、Po、Rn和Ac等。它们在周期表中的正确位置很容易判断出来。此外，研究者短时间内还发现了30多种明显是新元素的元素种类，它们被命名为钍放射物、镭放射物、锕X、铀X、钍X等，以此说明它们大概是由哪种元素得到的。X表示未知的种类，后来知道它们大多是不同元素的同位素。例如，铀X之后被确认为钍的同位素。

81

一些周期表的设计者如范登布鲁克等，尝试将这些新“元素”纳入我们上面所见的扩展的周期表中。同时，两位瑞典人丹尼尔·斯特伦霍尔姆和特奥多尔·斯韦德贝里制作的周期表，则强行将这些奇异新品种中的一部分放在同一个位置。例如，他们在惰性气体氙下方填入了镭放射物、锕放射物和钍放射物。这似乎表明他们预想到了同位素，但没有认清这种现象。

在门捷列夫去世的1907年，美国辐射化学家赫伯特·麦科伊得出结论：“放射钍在化学过程中与钍完全不可区分。”这个观察非常重要，很快就在其他许多对之前认为是新元素的物质中一再出现。弗雷德里克·索迪对这一观察做出了完整的评判，他也曾经是卢瑟福的学生。

对索迪而言，化学不可区分性只代表一件事，即这些元素是同一化学元素的两种或更多种形式。1913年，他发明了“同位素”这个词，来表示同一元素在化学上完全不可区分，但具有不同原子量的两种或更多种原子。化学不可区分性还由弗雷德里希·帕内特和格奥尔格·冯·赫维西在铅和“放射铅”的情况中注意到了。卢瑟福曾让他们用化学方法分离两者。在用20种不同的化学方法尝试完成这项艰巨任务之后，他们不得不承认自己完全失败了。这在某些方面是失败，但也进一步加强了这一认识：一种元素（在这里是铅）可以存在化学上不可分辨的同位素。这并不是白费工夫，因为帕内特和冯·赫维西的努力使他们建立了一套新的技术——分子的放射性标记法，这为一门在生物化学和制药研究等领域的应用中具有深远影响的分支学科奠定了基础。

82

1914年，哈佛的T. W.理查兹的研究为同位素提供了更有力的支持。他测量了同一元素两种同位素的原子量。他也选了铅，因为这种元素有许多放射性衰变系列。毫无意外，由不同过程形成的铅原子，涉及非常不同的中间元素，最后得到的铅原子的原子量差了0.75个单位之多。之后，这个结果由其他人增大到0.85个单位。

最后，同位素的发现进一步澄清了调换对现象，诸如一直困扰门捷列夫的碲和碘。尽管碲在周期表中排在碘之前，但它的原子量比碘更大，因为碲的所有同位素的平均原子量偶然地比碘的同位素的平均原子量更大。因此研究者就把原子量看成一种由元素所有同位素的相对丰度决定的偶然量了。就元素周期表而言，更基本的量是范登布鲁克-莫塞莱的原子序数，或按后来所知的应叫核内质子数。一种元素的身份是由它的原子序数而不是原子量得到的，因为原子量会依元素分离自不同样本而发生变化。

83

尽管碲的平均原子量比碘更大，但它的原子序数要小一个单位。如果用原子序数代替原子量作为元素的排序原则，碲和碘都可以放入符合它们化学行为的族里。因此，这说明之所以会有调换对，是因为在20世纪初之前所有的周期表都采用了错误的排序原则。

84

第七章

电子结构

第六章中讨论的大部分内容是经典物理中的发现，不需要量子理论。对X射线和放射性来说确实如此，其研究大体上不涉及任何量子概念，尽管之后的研究者会用量子理论澄清某些方面的事实。此外，第六章讨论的物理也大多是产生原子核的过程。放射本质上是核裂变，元素嬗变同样也发生在原子核的层面。并且，原子序数是原子核的性质，同位素靠同种元素不同的原子质量区分，而原子质量基本上就是原子核的质量。

在本章中，我们将探索原子中电子的相关发现，这项研究伊始就需要用上量子理论。不过首先，我们要谈一谈量子理论本身的发展。量子力学始于20世纪之前的德国，那里有一批物理学家尝试着理解内壁涂黑的小空室中的辐射行为。研究者将不同温度下这种“黑体辐射”的光谱行为仔细记录下来，再从数学上建立所得图像的模型。这个问题很长时间都没有得到解决，直到普朗克在1900年提出一个大胆的猜想，认为辐射的能量是 85
由离散的包或“量子”构成的。普朗克本人似乎一直不大情愿接受他的新量子理论的重大意义，于是如何应用它就留给其他人了。

量子理论断言能量以离散的包传递。阿尔伯特·爱因斯坦在1905年用这个理论成功地解释了光电效应。（他或许是20世纪最了不起的科学家。）他的研究结论是，可以认为光具有一种量化的，或者说粒子的属性。然而，爱因斯坦很快就将量子力学看成一套不完整的理论，终其余生对它持批评态度。

1913年，丹麦人尼尔斯·玻尔将量子理论应用于氢原子，他像卢瑟福一样猜测，氢原子由一个中心原子核和环绕核的电子组成。玻尔假设电子能够具有的能量只以离散值出现，用图像化的语言来说，就是电子能够存在于原子核周围任意数量的壳层或轨道上。这个模型可以在一定程度上解释氢原子，实际上还有任何元素的原子的两个行为特征。第一，它解释了氢原子在被突然施与电能时为什么会产生不连续的光谱，且其中只能观测到特定的频率。玻尔推理道，这种行为会在电子从一个可行的能级跃迁到另一个时出现。这种跃迁伴随着特定能量的释放或吸收，精准对应于原子两个能级之间的能量差。

第二，模型差强人意地解释了为何电子不会损失能量并撞入原子核，而经典力学预测做圆周运动的带电粒子会出现这种行为。玻尔的回答是，只要电子还在固定轨道上，它们就不会损失能量。他还假设存在最低的能级，那里的电子无法再向更低能级做任何跃迁。

玻尔随后不再只限于氢原子，而将模型推广到任意多电子原子。首先，他假设原子序数为Z的中性原子有Z个电子。然后，他再着手建立任意原子里电子的排列方法。这样从单电子跳到多电子在理论上是否有效，还是存疑的，但这并不妨碍玻尔不顾一切地前进。他得到的电子构型如图23所示。

但玻尔指定电子在某个壳层的假设并没有数学依据，也没

1	H	1				
2	He	2				
3	Li	2	1			
4	Be	2	2			
5	B	2	3			
6	C	2	4			
7	N	4	3			
8	O	4	2	2		
9	F	4	4	1		
10	Ne	8	2			
11	Na	8	2	1		
12	Mg	8	2	2		
13	Al	8	2	3		
14	Si	8	2	4		
15	P	8	4	3		
16	S	8	4	2	2	
17	Cl	8	4	4	1	
18	Ar	8	8	2		
19	K	8	8	2	1	
20	Ca	8	8	2	2	
21	Sc	8	8	2	3	
22	Ti	8	8	2	4	
23	V	8	8	4	3	
24	Cr	8	8	2	2	2

图23　玻尔1913年编制的最初的原子电子构型

有借助任何量子理论的具体内容。他反而求助于化学证据，比如硼元素的原子像硼族其他元素的原子一样，可以形成三个键等。为此，硼原子就必须有三个外层电子。但即使靠的是这样一种初步和非演绎的理论，玻尔依旧对锂、钠、钾元素为什么出现在周期表的同一族，还有周期表中其他类似的关系等给出了第一个成功的电子层面的解释。在锂、钠、钾的例子中，这是因为这些原子的最外电子层都只有一个电子。

玻尔的理论有一些其他的局限。其中一个是它只能严格地应用于氢原子或He^{+}、Li^{2+}、Be^{3+}等单电子原子。研究者还发现这些“类氢”光谱中的一些谱线分裂成了意想之外的谱线对。德国的阿诺德·索末菲认为原子核可能处在一个椭圆焦点上，而

87 不是圆形的圆心上。他经过计算证明，玻尔的主电子壳层中还需要引入子壳层。玻尔的模型用一个量子数来标记每个离散壳层或轨道，而索末菲的模型需要两个量子数来确定电子的椭圆

88 轨道。玻尔借助这些新量子数，在1923年编制了一份更详细的电子构型系统，如图24所示。

几年之后，英国物理学家埃蒙德·斯托纳发现，还需要第三个量子数来确定氢原子和其他原子光谱中一些更精细的结构。随后在1924年，奥地利裔理论物理学家沃尔夫冈·泡利发现需要第四个量子数，用来标记电子具有的一种特殊的二值角动量。

89 这一新运动最后被称为电子“自旋”，尽管电子并不真的像地球那样在绕太阳做轨道运动的同时又绕地轴旋转。

这四个量子数彼此相关，有一套嵌套关系。第三个量子数的取值范围依赖第二个量子数的值，后者又依赖第一个量子数的值。泡利的第四个量子数有些不同，因为它可以取+1/2或

H	1				
He	2				
Li	2	1			
Be	2	2			
B	2	3			
C	2	4			
N	2	4	1		
O	2	4	2		
F	2	4	3		
Ne	2	4	4		
Na	2	4	4	1	
Mg	2	4	4	2	
Al	2	4	4	2	1
Si	2	4	4	4	
P	2	4	4	4	1
S	2	4	4	4	2
Cl	2	4	4	4	3
Ar	2	4	4	4	4

图24　玻尔1923年以两个量子数为基础编制的原子电子构型

−1/2，与其他三个量子数的值无关。第四个量子数的重要之处，就在于它的出现能给出一个很好的说明，来解释为什么从最靠近原子核的电子壳层算起，每个壳层都包含特定数量的电子（2、8、18、32个等）。

下面是这个构型的原理。第一个量子数n可以取从1开始的任意整数。第二个量子数用l标记，可以取下列与n有关的数值：

$$l = n-1, \cdots\cdots 0$$

例如，在$n=3$的情况下，l可以取2、1或0。第三个量子数标记为m_l，取值与第二个量子数有关：

$$m_l=-l,\ -(l-1),\cdots,0,\cdots,\ (l-1),l$$

例如，如果$l=2$，则m_l可以取的值是：

$$-2,\ -1,0,+1,+2$$

最后，第四个量子数标记为m_s，只可以取两个值，就是之前提过的+1/2或 −1/2单位的自旋角动量。因此四个量子数取值相关，

构成一个层级，可用来描述原子中任意的特定电子（图25）。

按照这个构型，为什么第三层（举个例子）总共可以容纳18个电子就很明显了。如果第一个量子数是壳层数3，那么第三

n	l的可能值	子壳层标记	m_l的可能值	子壳层轨道数	每个壳层电子数
1	0	1s	0	1	2
2	0	2s	0	1	
	1	2p	1, 0, −1	3	8
3	0	3s	0	1	
	1	3p	1, 0, −1	3	
	2	3d	2, 1, 0, −1, −2	5	18
4	0	4s	0	1	
	1	4p	1, 0, −1	3	
	2	4d	2, 1, 0, −1, −2	5	
	3	4f	3, 2, 1, 0, −1, −2, −3	7	32

图25　联合四个量子数解释每个壳层的总电子数

层就总共有2×3^2，即18个电子。第二个量子数可以取2、1或0。每个l值都产生数个可能的m_l，每个m_l又可以乘以因子2，因为第四个量子数可以取值+1/2或−1/2。

但第三个壳层可以容纳18个电子的事实并不能严谨地解释周期系统的某些周期为什么有18个位置。它也只在电子壳层严格按顺序排列的情况下才是一个严谨的解释。虽然电子壳层一开始是按顺序排列，但从第19号元素钾开始就不再如此了。我们从可以容纳2个电子的1s轨道开始构建电子构型，到2s轨道可以类似地容纳另外2个电子。然后是2p轨道，总共可以再容纳6个电子，以此类推。这个过程一直以可预测的方式继续，直到第18号元素氩，它的构型是：

$$1s^2, 2s^2, 2p^6, 3s^2, 3p^6$$

你或许会认为下一个元素，即第19号元素钾的构型是：

$$1s^2, 2s^2, 2p^6, 3s^2, 3p^6, 3d^1$$

其中最后一个电子占据了下一个标记为3d的子壳层。你会这么想，是因为这个模式到现在为止，一直是把一个区分电子加到与原子核距离递增的下一个可行的轨道上。不过，实验证据表明钾的构型应该是：

$$1s^2, 2s^2, 2p^6, 3s^2, 3p^6, 4s^1$$

第20号元素钙情况类似，新电子也会进入4s轨道。但下一个元素，第21号钪的构型一般又被认为是（见第八章）：

$$1s^2, 2s^2, 2p^6, 3s^2, 3p^6, 4s^2, 3d^1$$

这种连续元素中新填充的电子在可行轨道间来回跳跃的情况会出现许多次。我们按周期表顺序移动，对区分电子的总结见图26。

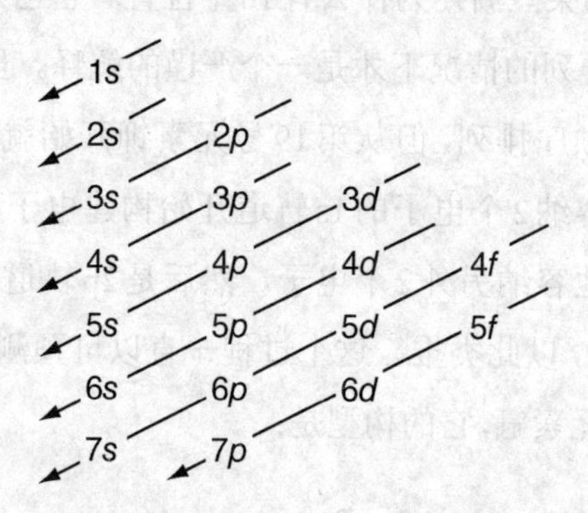

图26　轨道填充的大致顺序。更严格的区分电子顺序见第109页[①]

采取这种填充顺序的结果，就是周期表中连续周期包含的元素数量为2、8、8、18、18、32等，从而表现出除了第一个周期外，每个周期都“重复一次”的现象。

92

虽然四个量子数的组合规律给出了壳层在哪里结束的严谨解释，但它没有给出周期在哪里结束的同样严谨的解释。不过，通过这些填充顺序可以推导出一些结论，但这些推断多少依赖你要解释的事情。我们知道周期在哪里结束，是因为我们知道惰性气体是第2、第10、第18、第36、第54号元素等。同样，我们是通过观察，而不是通过理论知道轨道填充的顺序。教科书在解释周期表时很少会承认这样的结论：量子物理不能完全解释

① 指英文原书页码，见本书边码。下同。

周期表。目前还没有人从量子力学的原理推导出轨道的填充顺序。这并不是说我们以后也不能做到，也不是说电子填充顺序在某种意义上无法从根底用量子力学解释。

化学家与构型

J. J.汤姆生在1897年发现了电子，这件事既开启了全新的实验路线，也开启了各种各样物理上的新解释。汤姆生还是讨论电子以何种方式在原子中排列的先驱之一，不过他的理论并不十分成功，因为当时还不太清楚每个原子中存在多少电子。我们也看到了，这方面的第一个重要理论由玻尔提出，他也向原子领域和电子排布工作引进了能量量子化概念。玻尔成功发表了一套包含许多已知原子的电子构型系统，但这是在参考了多年来人们整理的化学和光谱行为之后取得的成果。

93

不过，这段时间化学家在干什么呢？相比玻尔和其他量子物理学家，他们对探索电子做出了哪些尝试？要展开这段调查，我们必须先回到电子刚被发现五年后的1902年。美国化学家G. N.刘易斯当时在菲律宾工作，他画出了图27中的草图（原稿保存至今）。图中，他假设电子处在立方体的角上，我们在周期表上每移至下一个元素，就会有一个新电子加到角上。现在看，选用立方体是相当奇怪的，因为我们知道电子绕核运动。但刘易斯的模型有一个很有用的地方，即它能与周期表联系起来。8是性质重复出现前元素必经的一个数。

刘易斯由此提出，化学周期性和各个元素的性质由围绕原子核的最外层电子立方体上的电子数量决定。虽然这个模型在认为电子呈静态这方面是完全错误的，但它的立方体选得自然

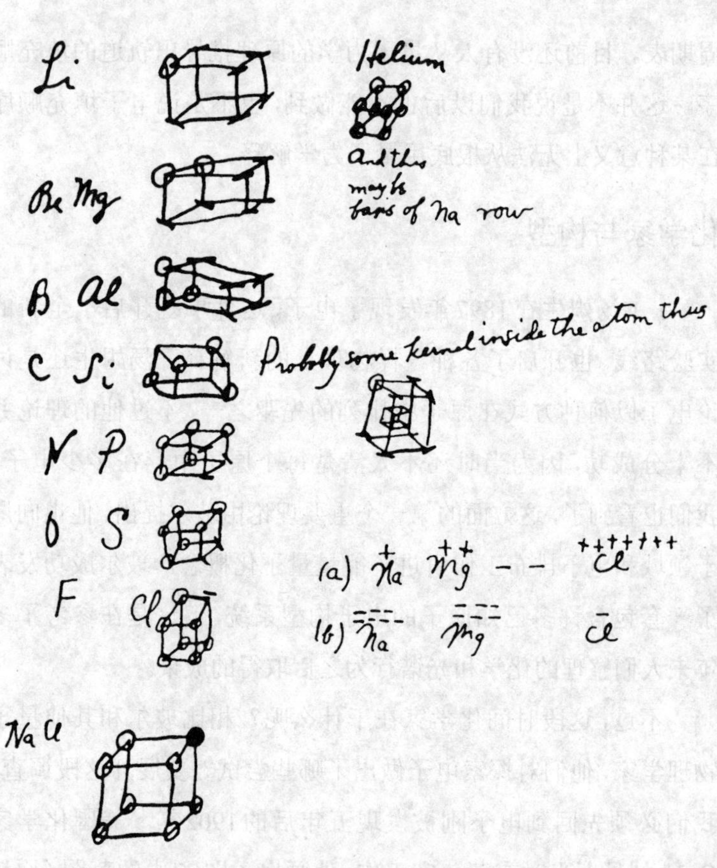

图27　刘易斯的原子立方体草图

94 而巧妙，反映了化学周期性基于八个元素间隔的事实。

在这些著名的草图中，刘易斯给出了钠原子和氯原子形成化合物的图解，其中一个电子从钠原子处转移到氯原子处，占据了氯原子外层立方体上没有电子的第八个角。刘易斯等待了14年才把这些想法发表出来，并将它们拓展到其他形式的化学键，也就是共价键。共价键涉及不同原子间共享电子，而不是电子转移。

刘易斯考虑了很多已知的化合物，他统计了它们所含原子的外层电子数，得出电子数在大部分情况下都是偶数的结论。95这个事实提示他化学键也许来自电子对，这个观念很快成为整个化学的中心，即便经历了接下来的关于化学键的量子力学理论，它在本质上至今仍然正确。

为了表示两个原子共享电子，刘易斯画了两个相连的立方体，它们共用一条棱，即两个电子。同样，双键以两个共用一个面，即四个电子的立方体表示（图28）。但随后就有问题了。在有机化学中，我们已经知道一些化合物，比如乙炔C_2H_2，包含三键。刘易斯意识到，用立方体角上电子这一模型无法表示三键。在同一篇论文中，他换了一个新模型，其中四对电子位于四面体（不再是立方体）的角上。三键是两个四面体共用一个面形成的。

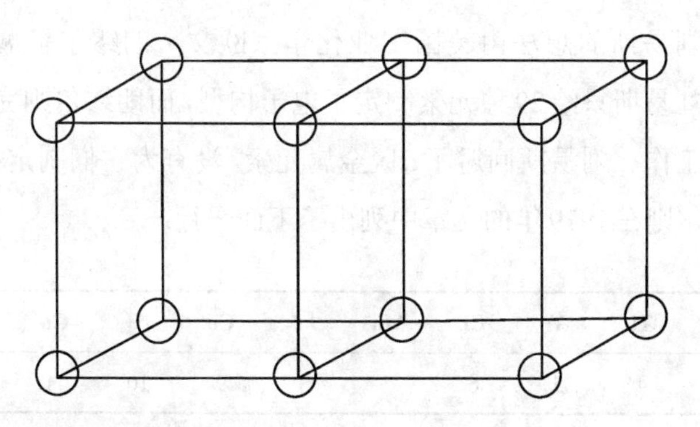

图28 刘易斯对两原子间双键的图示

刘易斯在这篇论文中还回到了原子电子构型的问题，现在，他将他的系统扩展到29种元素，如图29所示。在介绍下一位贡献者之前，我们值得在这里多停留一会儿，讨论一下G. N.刘易斯这位20世纪大概最著名的化学家为什么没有获得诺贝尔奖。

96 一个原因是他在实验室中因氰化氢中毒英年早逝，而诺贝尔奖不会在身后追授。另外一个很重要的原因是，刘易斯在学术生涯中树敌过多，没有亲近的同事为他提名。

1	2	3	4	5	6	7
H						
Li	Be	B	C	N	O	F
Na	Mg	Al	Si	P	S	Cl
K	Ca	Sc		As	Se	Br
Rb	Sr			Sb	Te	I
CS	Ba			Bi		

图29　刘易斯列出的29种元素的外层电子结构。每列第一行的数字表示每种原子的最外层电子数(作者整理)

刘易斯的想法由美国工业化学家欧文·朗缪尔扩展并推广。刘易斯只给29种元素制定了电子构型，而朗缪尔则完成了全部工作。刘易斯回避了d区金属元素，没有为它们制定构型，朗缪尔则在1919年的文章中列出了下面一行：

Sc	Ti	V	Cr	Mn	Fe	Co	Ni	Cu	Zn
3	4	5	6	7	8	9	10	11	12

朗缪尔像之前刘易斯所做的一样，用元素的化学性质指导自己的构型，而不是运用量子理论(图30)。毫无意外，这些化学家97 也可以改进电子构型，就像玻尔等物理学家所做的探索一样。

1921年，英国化学家查尔斯·伯里在位于阿伯里斯特维斯的威尔士大学学院工作，他对刘易斯和朗缪尔工作背后的一个

TABLE I.

Classification of the Elements According to the Arrangement of Their Electrons.

Layer.	N	E = 0	1	2	3	4	5	6	7	8	9	10
I			H	He								
II*a*	2	He	Li	Be	B	C	N	O	F	Ne		
II*b*	10	Ne	Na	Mg	Al	Si	P	S	Cl	A		
III*a*	18	A	K	Ca	Sc	Ti	V	Cr	Mn	Fe	Co	Ni
			11	12	13	14	15	16	17	18		
III*a*	28	Niβ	Cu	Zn	Ga	Ge	As	Se	Br	Kr		
III*b*	36	Kr	Rb	Sr	Y	Zr	Cb	Mo	43	Ru	Rh	Pd
			11	12	13	14	15	16	17	18		
III*b*	46	Pdβ	Ag	Cd	In	Sn	Sb	Te	I	Xe		
IV*a*	54	Xe	Cs	Ba	La	Ce	Pr	Nd	61	Sa	Eu	Gd
			11	12	13	14	15	16	17	18		
IV*a*			Tb	Ho	Dy	Er	Tm	Tm_2	Yb	Lu		
		14	15	16	17	18	19	20	21	22	23	24
IV*a*	68	Erβ	Tmβ	$Tm_2β$	Ybβ	Luβ	Ta	W	75	Os	Ir	Pt
			25	26	27	28	29	30	31	32		
IV*a*	78	Ptβ	Au	Hg	Tl	Pb	Bi	RaF	85	Nt		
IV*b*	86	Nt	87	Ra	Ac	Th	Ux_2	U				

图30　朗缪尔的周期表

理念提出了质疑。这个理念就是假设在遍历周期表、每次为下一个新元素添加一个电子时，电子壳层总是按顺序被填充。伯里称他修正的电子构型更符合已知化学事实。

总而言之，物理学家对人们尝试理解周期表的基础起到了巨大的促进作用，而当时的化学家在许多情况下都可以应用新的物理观念（比如电子构型）来得到更好的效果。

98

第八章

量子力学

第七章谈的是早期的，尤其是玻尔（图31）的量子理论对解释周期表的影响。这一解释因泡利的贡献达到顶点，他引入了第四个量子数和以他的名字命名的不相容原理。之后，我们才能够解释为什么围绕原子的壳层会容纳特定数量的电子。（第一层2个，第二层8个，第三层18个，等等。）如果你接受填充这些壳层里的轨道的正确顺序，就可以解释周期长度为何实际上是2、8、8、18、18等。但任何对周期表有价值的解释，都应该能从第一原理得到这一列数值，而**无须**接受观测到的轨道填充顺序。

玻尔的量子理论，即便有了泡利的贡献助力，也只是一种更先进理论的基石。玻尔-泡利版的理论通常被称为量子理论，有时候也叫旧量子理论，以区别于之后在1925年到1926年发展起来并取而代之的量子力学。用“理论”这个词来称呼旧版本不太妥当，因为它加深了人们对“理论”的误解，以为理论是一种模糊的、尚未成为科学定律或其他稳固知识的半成品。

但在科学中，理论是指得到高度实证（尽管不可能证实）的

图31 尼尔斯·玻尔

知识体，比科学定律地位更高。许多不同的定律往往被纳入一套包罗万象的"理论"。所以量子力学这套后续的理论，完完全全是与玻尔的旧量子理论一样的"理论"，尽管它的应用更广泛，也更成功。

玻尔的旧量子理论有许多不足，比如它不能解释化学键。不过，在量子力学到来后，这一切都变了。这时，我们就可以比G. N.刘易斯的流浪汉小说式的观念（成键只是分子中两个或多个原子之间共享电子）更进一步。根据量子力学，电子的行为既像粒子，也像波。奥地利人埃尔温·薛定谔写下绕核电子运动 100

的波动方程，由是迈出了决定性的一步。薛定谔方程的解，代表了在原子中找到电子时它们可能处于的量子状态。不久之后，亨德和密立根两位物理学家各自提出了分子轨道理论，其中，成键是分子中每个原子的电子波相消干涉的结果。

但我们要回到原子和周期表。根据玻尔的理论，我们能够计算能级的只有那些仅有一个电子的原子，比如氢原子和He^{+1}、Li^{+2}、Be^{+3}等单电子离子。在多电子原子中（“多”指的是多于1），玻尔的旧量子理论就无能为力了。反过来，用新的量子力学，理论家就能应对多电子原子，虽说必须承认，我们只能以近似而非精确的方式应对。这是因为有数学限制。任何一个电子数多于1的系统都有一个所谓的“多体问题”，而这些问题只有近似解。

所以任意多电子原子量子态的能量都可以由第一原理求出，尽管它们与测量到的能量非常吻合，但仍是近似的。而周期表的一些全局特征至今还是无法由第一原理得到。之前说过的轨道填充顺序就是其中一例。

任何原子中构成子壳层和壳层的原子轨道，都是按照标示特定轨道的前两个量子数之和以递增的顺序填充的。这一现象可以总结为$n + l$法则或马德隆法则（以埃尔温·马德隆命名）。$n + l$的值从对应1s轨道的1开始逐渐增大，依此顺序填充轨道。

从1s轨道起始的斜箭头出发，然后移动到下一组斜线，就得到了如下轨道填充顺序（图26）：

$$1s < 2s < 2p < 3s < 3p < 4s < 3d < 4p < 5s \text{ 等}$$

虽然量子力学有这些耀眼的成就，但它仍然无法用纯理论的方式得到$n + l$值的序列。这不是贬低新理论的成功，它的确

101

让人叹为观止。我只是说还有需要阐明的地方。也许未来的量子物理学家可以成功导出马德隆法则，也许他们需要更有用的未来理论才能导出。我并非暗示对量子力学而言化学现象存在某种固有的、“鬼魅的”不可简化性，而是关注迄今为止我们在周期表语境中实际做到了什么。

不过，让我们再回到量子力学对化学周期性的解释上来。之前已经提过，我们可以写出周期表中任何原子系统的薛定谔方程并求解，而不需要输入任何实验值。例如，物理学家和理论化学家已成功计算出了周期表中各原子的电离能。拿计算结果与实验值比较，可以发现两者符合程度相当高（图32）。

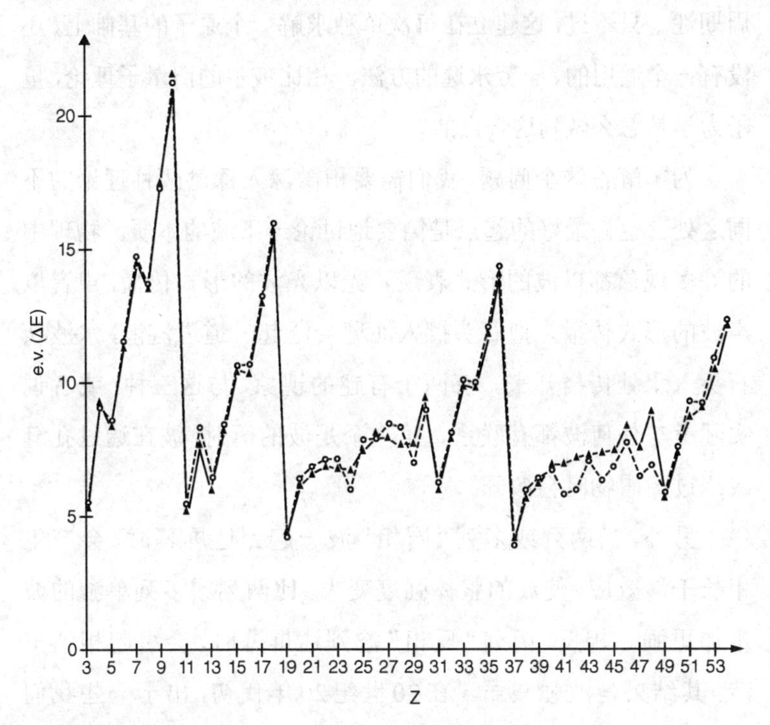

图32　第1到第53号元素的理论计算电离能（三角形）和观测电离能（弧形）

102 电离能正好是原子明显表现出周期性的一种性质。我们从Z=1的氢原子开始增大原子序数，到下一个元素氦，电离能也增大。到Z=3的锂，电离能急剧减小。随后电离能大体上又开始增大，直到化学上类似于氦的氖元素（氖也是一种惰性气体）。这种电离能整体上升的模式，在周期表的各个周期不断重复。每个周期里的最小值都是第1族的元素锂、钠、钾等，最大值则像上面看到的，是氦、氖、氩、氪、氙这些惰性气体。图32显示了将理论计算值连起来后的折线。103

结论是，即便量子力学还不能大致得到求出轨道填充顺序的方程（$n + l$法则），它仍解释了蕴藏在所有元素原子性质里的周期性。只不过，这建立在每次单独求解一个原子的基础上，还没有一个通用的、一劳永逸的方法。相比玻尔的旧量子理论，量子力学是怎么做到这一点的？

为了解答这个问题，我们需要稍微深入探讨两种理论的不同之处。也许最好的起点是简要地讨论一下波的本质。物理中的许多现象都以波的形式表现。光以光波的形式传播，声音以声波的形式传播。把石头掷入池塘会泛起一道道涟漪，水波从石头入水处传播开来。有两个有趣的现象，与这三种，或者说实际上与任何波都有联系。第一个是波的衍射，波在通过孔缝或绕过障碍物时会散开。

另外，当两列或多列"同相"波一起到达屏幕时，会产生相长干涉效应，使波的整体强度变大，比两列或多列单独的波相加更强。相反，两列"反相"波到达屏幕时，会引起相消干涉，其结果是波强减弱。在20世纪20年代初，出于一些我们在这里不便细说的原因，研究者猜测电子这类粒子在某些条件下或许也会表现得像波。这个观点简单点说，逻辑与爱因斯坦解释光电效应的逻辑相反，即把“光波的行为也像粒子”反过来。

为了测试电子这类粒子是不是真的像波一样，我们必须看看电子能否也发生其他类型的波能够发生的衍射和干涉效应。104 或许出人意料的是，这些实验都成功了，电子的波动性从此得到确认。此外，径直打向金属镍单晶体的一束电子还产生了一系列同心环，这说明电子波在绕过晶体散开时形成了衍射图样，并不是直接从表面反弹回来。

从这一点出发，电子和其他基本粒子都应该具有一种分裂的本质，既表现为粒子也表现为波。更具体地说，电子的波动行为的消息很快传到了更大的理论物理学圈子，埃尔温·薛定谔就在其中。这位奥地利理论学家，着手用过时的模拟波的数学工具来描述氢原子中电子的行为。他的假设是：一、电子的行为像波；二、电子的势能包括它受到原子核吸引的能量。薛定谔以数学物理学家一直用来解这类偏微分方程的方法，利用边界条件解出了他的方程。

设想一根固定的吉他弦，它的两端分别为上弦枕和琴桥。现在拨弦，弹出一个空弦音。如图33所示，弦会以半波长的形态沿上下方向振动。不过弦还有其他的振动方式吗？答案是，它还能够以2个半波长（即1个完整波长）的形态振动。105 手指轻轻按在指板中心（12品处），在另一只手拨动响孔处的弦的瞬间松手，就可以产生这种振动。如果恰到好处，发出的声音就是让人愉悦的钟鸣声，音乐上称为泛音。弦可以振动的其他模式还有3、4、5个半波长等。弦只能以半波长的整数倍振动。例如，不可

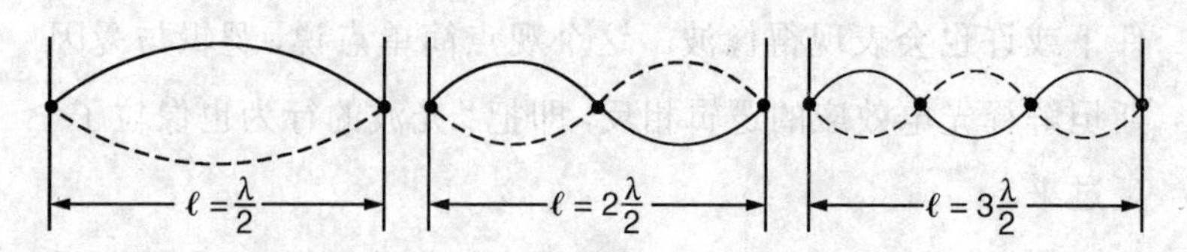

图33　边界条件和量子化的表现

能有某种模式以$2\frac{1}{2}$个半波长或$3\frac{1}{3}$个半波长振动。

用数学与物理术语来描述就是，强制的边界条件（在吉他这个例子中是两个固定端点）会引起一系列以整数为特征的运动。这无非相当于量子化，或者说是限制为某个值的整数倍。薛定谔运用类比的方法，将这种数学边界条件应用到他的方程中，算出的电子能量就量子化了，这正是玻尔之前强行向他的方程中引入的条件。薛定谔的做法有本质上的改进，这让他成功得到量子化的能量，而无须人工引入。这就是深度处理的标志，量子化的能量是理论自然而然的一部分。

还有另一个区分新量子力学和玻尔的旧量子理论的重要因素。这第二个因素是德国物理学家沃纳·海森堡发现的，他找到了粒子位置不确定性乘以其动量不确定性的一个非常简单的关系：

106

$$\Delta x \cdot \Delta p \geqslant h/4\pi$$

这个方程暗示我们必须抛弃一个常识：像电子这样的粒子同时具有确定的位置和动量。海森堡的关系认为，我们越确定电子的位置，就越不能确定它的动量，反之亦然。粒子的运动仿佛带有一种模糊的、不确定的本质。而玻尔模型处理的是电子精确的行星式轨道，它们绕原子核旋转；量子力学中的新观点则是我们或许不能再讨论确定的电子轨道了。反过来，理论只能退而求其次，用概率（不确定性）代替确定的语言。这个观点和电子是波的观点结合后，呈现的图像就与玻尔的模型非常不同了。

量子力学认为电子是弥散在壳层中的。仿佛我们熟知的那种粒子变成了气体，飘散着充满整个壳层，其范围大致对应于玻尔二维轨道在三维中的呈现形式。此外，由于海森堡不确定性，量子力学中的壳层电荷没有明确的边缘。我们之前讨论过的第一轨道，即1s轨道，其实是一个空间，我们有90%的概率在其中找到一般被看作局域性粒子的电子。这只是薛定谔波动方程的第一个解。随着我们远离原子核，更大的轨道上开始出现不同于球壳的形状（图34）。

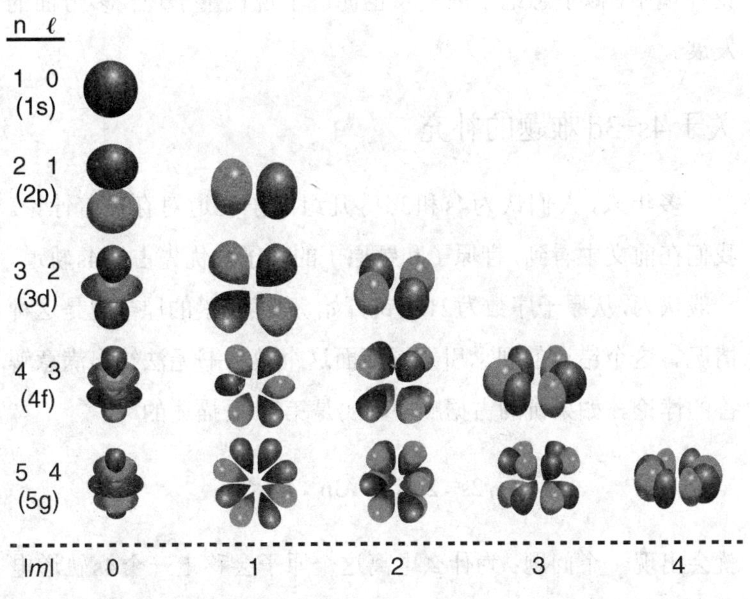

图34 s、p、d、f、g轨道

现在来考虑有关量子理论、量子力学和周期表的更广阔的图景。我们在第七章中知道，玻尔一开始是在氢原子的原子结构研究中引入了量子概念。但他在同一批论文里，就已经假设 107 能量量子化也会出现在多电子原子中，以此解释周期表的形式。紧接着玻尔的第一个量子数，进一步引入三个量子数，也是为了更好地解释周期表。甚至在玻尔闯进量子概念世界之前，当时顶尖的原子物理学家J. J.汤姆生就已在推测各个原子中的电子排布来编排周期表了。

因此，周期表成了原子物理理论、量子理论和之后量子力学早期几个研究方向的测试台。今天，人们认为化学，尤其是周期表，可以用量子力学完全解释清楚。即使情况绝非如此，这个理论依旧扮演的解释角色也是无法否认的。但人们在现在的还原 108 论环境中，似乎忘记了周期表也促进了现代量子理论多方面的发展。

关于4s-3d难题的补充

多年来，人们认为4s和3d等几组原子轨道对存在着悖论。我们在前文中看到，钾原子和钙原子的电子会优先占据4s轨道。一般认为，从原子序数为21的钪开始，后续元素的原子也是这种情况。这个错误的观念引发了下面这个教科书无法给出满意解答的悖论。如果钪的占据顺序确实是第92页描述的

$$1s^2, 2s^2, 2p^6, 3s^2, 3p^6, 4s^2, 3d^1$$

就会出现一个问题：为什么电离这个原子会移走一个4s轨道电子，而不是3d轨道电子？这个问题最近解决了，在本书第一版出

版之后。

实际上，钪的正确轨道占据顺序是：

$$1s^2, 2s^2, 2p^6, 3s^2, 3p^6, 3d^1, 4s^2$$

因此，优先电离4s电子不会再引起任何问题。我把前一种错误的想法称为“草率的构造”，认为钪和其后过渡金属的原子构型就只是在前一个原子的构型上再添加一个电子，在钙那里，优先占据的确实是4s轨道。现在人们认识到情况并非如此，而这确实是先前人们认为这个难题存在的原因。（见“延伸阅读”） 109

反常构型和最近的第二个发现

另一个先前与原子构型和周期表有关的神秘问题最近也得到了解决。它和我们沿周期表移动时，看到的20种过渡金属原子的不规则或者说反常行为有关。在第一过渡系中，铬原子和铜原子最外侧的s轨道上只有一个电子，而这一系的其余8种元素则有两个这样的电子。在第二过渡系中，10种元素中有6种表现出类似的反常。

如今，理论家欧根·施瓦茨对这些事实给出了令人满意的解释。他证明，如果对恰当的光谱能级做合适的加权平均，那么此时就可以认为，通常讨论的反常现象反映了这些原子的s^1和s^2构型是相对稳定的。从这个角度看，反常构型其实并不存在，只是各种构型都在争做能量最低的构型而已。（见“延伸阅读”） 110

第九章

现代炼金术：从空缺的元素到合成元素

周期表包含了大约90种自然存在的元素，到第92号元素铀为止。我说大约90种，是因为像锝这样的一两种元素是先由人工合成，然后人们才发现其在地球上自然存在。

化学家和物理学家相继合成了氢（1）和铀（92）之间空缺的元素。此外，他们还合成了25种以上铀之后的新元素，虽说后来人们发现其中的几种（比如镎和钚）在自然中也极微量地存在。

直至第118号（含）的所有元素都由人工合成了。最新加入的有第113、第115、第117号和第118号元素。它们的名称分别为鉨、镆、硇和鿫。

许多元素在合成时，都是先从特定的核开始，用小粒子轰击它，以增加原子序数，由此改变目标核子本身。最近，合成方法为用原子量相当的核子相撞，但目的依旧是形成更大、更重的核子。

我们有一种根本认识，即这些合成都源自一项重要的实验，这项实验由卢瑟福和索迪于1919年在曼彻斯特大学主持。卢瑟福和同事所做的就是用α粒子（氦核）轰击氮原子核，结果氮原子核转变成了另一种元素的原子核。反应还产生了一种氧的同位素，不过他们一开始并没有意识到这一点。卢瑟福第一个完成了将一种元素完全嬗变成另一种的实验。古代炼金术士的梦想成为现实，而这个普遍过程持续产生新的元素，直到今天。

不过，卢瑟福的反应并没有产生新元素，只是产生了某种既有元素的不常见的同位素。卢瑟福使用的α粒子来自另一种不稳定核子（如铀核）的放射性衰变。研究者很快发现，用氮原子以外的原子做靶原子，也可以发生类似的嬗变，但这个方法只适用于原子序数为20的钙之前的元素。如果要转化出更重的核，就需要比自然产生的α粒子能量更大的入射粒子，因为入射粒子必须克服高电荷靶原子的排斥。

20世纪30年代，加州大学伯克利分校的欧内斯特·劳伦斯发明了回旋加速器，情况发生了改变。这种机器可以将α粒子加速到自然α粒子速度的几百甚至上千倍。此外，另一种入射粒子——中子——也在1932年被发现了。它具有携带电荷为零的额外优势，这意味着它能够穿透靶原子，而不受到核内带电质子的任何排斥。

112

空缺的元素

20世纪30年代中期，周期表还有四个空位有待填补。这些元素的原子序数是43、61、85和87。有意思的是，其中三种元素已经由门捷列夫在许多年前明确地预测了，他称它们为类锰（43），类碘（85）和类铯（87）。这四种空缺元素中的三种是20

世纪先由人工合成出来而被发现的。

1937年，研究者在伯克利用回旋加速器做实验，他们用氘（氢的同位素之一，其原子量是丰度更高的氕的两倍）粒子流轰击一块钼靶。埃米利奥·塞格雷是其中一名研究者，他是从西西里来的博士后，当时正要把被辐射的金属板带回祖国。在巴勒莫，塞格雷和佩里埃分析了这些金属板之后，确信他们合成了一种新元素，原子序数为43，他们将其命名为锝（technetium）。

这是由嬗变得到的第一个全新元素，距卢瑟福证明这种可能性的经典实验已经过去了18年。研究者之后也找到了新元素锝在自然中存在的迹象，存量非常低。

第85号元素，即门捷列夫的类碘，是第二个由人工合成出来而被发现的。考虑到其原子序数为85，它应当可以从钋（84）或铋（83）合成。钋非常不稳定，而且有放射性，于是研究者将注意力转向铋。考虑到铋的原子序数比第85号元素小两个单位，轰击粒子显然还得是α粒子。1940年，达莱·科森、麦肯齐和塞格雷（当时已永久定居美国）在伯克利主持了这项实验，并得到了第85号元素的一种同位素，半衰期为8.3小时。他们将其命名为砹（astatine），取自希腊语astatos，意思是“不稳定”。人工合成的第三种空缺元素是第61号元素，它也是在伯克利的回旋加速器里得到的，这支实验队伍由雅各布·马林斯基、劳伦斯·格兰德宁和查尔斯·科里尔组成。反应是用氘原子轰击钕原子。

我们把这个故事讲完，再说第四种空缺元素，它是1939年由法国化学家玛格丽特·佩雷发现的。她最初是居里夫人的实验助手，发现新元素之时甚至还没有取得本科学位。她以祖国之名将新发现的元素命名为钫（francium）。这种元素不是人工合成的，而是作为锕元素自然的放射性衰变的副产品被发现的。佩雷最终升任教授，并成为法国最大的核化学研究机构的主任。

超铀元素

我们最后来看看第1到第92号之后的元素合成情况。1934年，即合成锝之前三年，罗马的恩里克·费米用中子轰击靶元素，以期合成超铀元素。费米认为他得到了两种元素，随即分别命名为“ausonium”（93）和“hesperium”（94）。但它们并非新元素。

同年，他在诺贝尔奖获奖演讲中宣布了自己的发现，但很快在之后发表纸质演讲稿时又撤销了宣布。一年后，1938年，奥托·哈恩、弗里茨·施特拉斯曼和莉泽·迈特纳发现了核裂变，这为费米的错误发现做出了解释。过程清楚了：诸如铀核的重核在与中子碰撞时，会分裂成两个中等大小的核子，而不会变成更大的核。例如，铀会生成铯和铷。

费米与合作者先前发现的是这样的核裂变产物，而不是他们一开始认为的重核核子。

真正的超铀元素

真正的第93号元素，最终在1939年由埃德温·麦克米伦及其合作者在伯克利确认。他们将其命名为镎（neptunium），因为它在周期表上位于铀（uranium）之后，正如海王星（Neptune）到太阳的位置排在天王星（Uranus）之后。团队中的化学家菲利

114

普·埃布尔森发现，第93号元素与根据周期表位置预测的类铼并不相似。正是基于这一点，加上之后对第94号元素钚的类似发现，格伦·西博格对周期表提出了一个重大修正。（见第一章）于是，从锕（89）开始的元素不再被视为过渡金属，它们类似于镧系金属。因此，第93和第94号等元素不必表现出类铼和类锇的性质，因为在修正的周期表中它们被移到了另外的位置。

第94到第97号元素钚、镅、锔、锫都在之后的20世纪40年代合成出来了。第98号元素锎发现于1950年。但这一串成果似乎要终结了，因为一般来说，原子核越重就越不稳定。问题变成我们需要积累足够的靶材料，才能指望用中子轰击它们，将它们转变为更重的元素。就在此时，转机出现了。1952年，太平洋上的马绍尔群岛附近进行了一次代号为“麦克”的热核试爆。

其中一个结果是，核爆产生了强度很高的中子流，这让当时用其他方法都无法发生的反应可能发生了。例如，铀238可以与15个中子碰撞形成铀253，然后丢失7个β粒子，最后形成第99号元素锿。

$$^{238}_{92}U + 15\,^{1}_{0}n \rightarrow {}^{253}_{92}U \rightarrow {}^{253}_{99}Es + 7\,^{0}_{-1}\beta$$

第100号元素镄以类似的方式得到，是同一次爆炸中高能中子流的产物。研究者在分析了附近太平洋海岛上的土壤之后发现了它。

从101到106

按顺序进一步迈向更重的核，需要一种非常不同的方法，因为Z=100之后的元素不会发生β衰变。许多技术创新出现了，比如不再用回旋加速器而改用直线加速器，研究者因此可以加速高强度的离子，使其具有确定的能量。此时入射粒子也可以比中子和α粒子更重。在冷战时期，只有美国和苏联这两个超级大国具备这类设施。

1955年，研究者在伯克利用直线加速器以下面的方法得到了第101号元素钔：

$$^{4}_{2}He + ^{253}_{99}Es \rightarrow ^{256}_{101}Md + ^{1}_{0}n$$

核子组合的可能性更多。例如，研究者在伯克利以下面的反应制出了第104号元素𬬻：

$$^{12}_{6}C + ^{249}_{98}Cf \rightarrow ^{257}_{104}Rf + 4\,^{1}_{0}n$$

在俄国杜布纳，研究者以下面的反应得到了这种元素的另一种同位素：

$$^{22}_{10}Ne + ^{242}_{94}Pu \rightarrow ^{259}_{104}Rf + 5\,^{1}_{0}n$$

第101到第106号的全部6种元素都以这种方法合成。部分原因在于冷战的紧张局势，美苏就其中多种元素的合成都爆发了激烈的争论，而且争论持续了很多年。但到第106号元素之后，新问题出现了，我们又需要新方法了。德国科学家此时进入战局——他们在达姆施塔特建立了重离子研究所。新方法名为"冷核聚变"，不过它与1989年化学家马丁·弗莱施曼和斯坦利·庞斯错误宣布的试管冷核聚变毫无关系。

超铀冷核聚变是让核子与另一个核子以较慢速度（相较之前的速度而言）相撞的技术。这样产生的能量更低，因此降低了合成核子分裂的概率。这项技术由尤里·奥加涅相首创，他本是苏联物理学家，但在德国大展拳脚。

第107号元素之后

20世纪80年代初，第107（铍）、第108（𬭶）和第109号（鿏）元素都由德国人用冷核聚变方法合成了。但此时又一道障碍变得明显。在柏林墙倒塌、苏联解体之后，美、德、俄开始合作，运用了很多新观念和新技术，但依旧一无所获。1994年，经历了十年停滞后，重离子研究所宣布用铅和镍离子碰撞合成了第110号元素。

117

他们得到的同位素的半衰期只有170毫秒。不出所料，德国人将此元素命名为“𫟼”（darmstadtium），因为之前美国和俄国的团队分别命名了“锫”（berkelium）和“𬭊”（dubnium）。一个月后，德国人得到了第111号元素，将它的名字定为“𬬭”（roentgenium），以纪念X射线的发现者伦琴（Röntgen）。他们在1996年2月合成了下一个，即第112号元素。2010年，该元素被正式命名为“鿔”（copernicium）。

第113到第118号元素

自1997年以来，研究者接连发布合成第113到第118号元素的消息，最后一个元素是2010年合成的第117号元素。考虑到质子为奇数个的核子总是比偶数个的核子更不稳定，这种情况便可以理解了。出现这种稳定性差异的原因是，质子具有半自旋，它们像电子那样两两进入能量轨道，自旋方向相反。由此，偶数个质子的核子形成的总自旋往往为零，因此比那些包含未配对质子的核子（就像第115和第117号元素这些含奇数个质子的核）更稳定。

研究者非常期待合成第114号元素，因为曾有预言认为那里是一座“稳定岛”的开始，也就是说表中这部分元素的核子具有更强的稳定性。首先宣布发现第114号元素的是杜布纳实验室，时间为1998年末。但直到1999年，研究者通过用钙48离子轰击钚靶等进一步实验，才明确得到这一元素。伯克利和达姆施塔特的实验室最近确认了这一发现。在本书写作时，研究者已经报告了大约80起涉及第114号元素的衰变，其中30起是从第116、第118号这类核子更重的元素衰变而来。第114号元素寿命最长的同位素的原子量为289，半衰期约为2.6秒，与这一元素具有更强稳定性的预期相符。

1998年12月30日，杜布纳实验室和利弗莫尔实验室联合发表了一篇论文，宣布在以下反应中发现了第118号元素：

$$^{86}_{36}Kr + ^{208}_{82}Pb \rightarrow ^{293}_{118}Og + ^{1}_{0}n$$

在日本、法国和德国的几次重复尝试都失败之后，发布方在2001年7月正式撤回了论文。之后出现了一些争议事件，比如解雇该研究团队最初宣布消息的那名资深成员。

两年后，杜布纳宣布了一些新消息；随后，加利福尼亚的劳伦斯·利弗莫尔实验室也在2006年发布了进一步的信息。美国和俄罗斯科学家联合发布了更有分量的公告，称他们在如下反应中又探测到了第118号元素的4种衰变：

$$^{48}_{20}Ca + ^{249}_{98}Cf \rightarrow ^{294}_{118}Og + 3^{1}_{0}n$$

研究者非常自信，认为结果是可靠的，因为他们预计探测结果是随机信号的可能性要小于十万分之一。不用说，研究者对这种元素还没做过化学实验，因为制造出来的原子非常少，而且它们的寿命非常短，不到1毫秒。

119

2010年，杜布纳的一个很大的研究团队，与美国一些实验室的研究者合成并描述了一种更不稳定的元素，即第117号元素石田。周期表到达了一个有趣的节点，自然存在和用特定实验人工得到的所有118种元素都集齐了。其中包含了铀元素之后的26种引人瞩目的元素，而我们一般认为铀是自然存在的最后一种元素。使第7周期得以完整结束的最后4种元素是鉨（113）、镆（115）、石田（117）和氭（118）。在本书写作之时，研究者还计划尝试合成第119和第120号等更重元素，我们没理由认为可合成元素的终点会很快到来。

合成元素的化学性质

超重元素的存在提出了一个有趣的新问题，对周期表构成了一个挑战。它还为理论预测与实验发现之间的竞争带来了一个有趣的新结合点。理论计算认为，随着原子核的电荷增加，相对论效应也会越发重要。例如，原子序数中等的第79号元素金，它的特征性金色现在就是用相对论解释的。核电荷越大，内壳层电子的运动速度就越大。由于获得了相对论性速度，这些内壳层电子被核吸引得更近，由此对最外层电子产生更大的屏蔽效应，而任何元素的化学性质都是由最外层电子决定的。据预测，一些原子的化学行为会与按它们在周期表上的位置推断出来的不一样。

因此，相对论效应就为测试周期表普适性带来了最后的挑战。许多年来，研究者一直在发表这样的理论预测，但在检测了第104和第105号元素（分别为铲和钳）的化学性质后，发表情况达到了高潮。研究者发现，铲和钳的化学行为实际上和按这两种元素在周期表上的位置所做的直觉推断大不相同。铲和钳似乎不像它们应该表现的那样，分别类似于铪和钽。 120

例如，1990年，伯克利的K. R.切尔文斯基发表论文称，第104号元素铲的溶液化学与排列在它之上的两种元素锆和铪都不同。同时，他还发文称铲的化学性质类似于表上位置离得很远的元素钚。至于钳，早期研究显示它的行为也不像它之上的元素钽（图35），反倒与锕系的镤元素有更高的相似度。在其他实验中，铲和钳都似乎更像铪和钽之上的两种元素，即锆和铌。

3	4	5	6	7	8	9	10	11	12
Sc	Ti	V	Cr	Mn	Fe	Co	Ni	Cu	Zn
Y	Zr	Nb	Mo	Tc	Ru	Rh	Pd	Ag	Cd
Lu	Hf	Ta	W	Re	Os	Ir	Pt	Au	Hg
Lr	Rf	Db	Sg	Bh	Hs	Mt	Ds	Rg	Cn

图35　包括第3到第12族的周期表片段

直到研究者检查了之后的𬭳元素（106）和𬭛元素（107），才发现它们表现出预期的周期行为。已发表论文的标题就能说明问题，比如《出奇普通的𬭳》《波澜不惊的𬭛》等。它们都提到一个事实：周期表一如往常。尽管相对论效应本应在这两种元素上体现得更显著，但预期的化学性质还是压倒了这一趋势。 121

𬭛表现得有如标准的第7族元素这个事实，可以从以下我之前提过的论证看出。它也代表了一类“闭环”，因为它包含一个三元素组。读者或许还记得第三章中谈到，发现三元素组是与同一族元素性质有关的数字规律的第一个提示。图36是锝、铼、铍与氧、氯组成的相似化合物的标准升华焓（固态直接转变为气态需要的能量）的测量数据。

TcO_3Cl = 49 kJ/mol
ReO_3Cl = 66 kJ/mol
BhO_3Cl = 89 kJ/mol

图36　第7族元素的升华焓表明第107号元素是该族名副其实的成员

用三元素组方法预测BhO_3Cl的值为83 kJ/mol，与上面的实验值89 kJ/mol误差只有6.7%。这个事实进一步支持了铍是名副其实的第7族元素的观点。相对论效应带给周期律的挑战在第112号元素鿔的例子中甚至更加明显。这次，相对论计算结果表明其化学行为会变化到惰性气体的程度，而不像周期表里位于它之上的汞。实验测量第112号元素的升华焓，表明这种元素确实属于锌、镉、汞所在的第12族，与之前的预测不同。

第114号元素的故事也类似。先期计算和实验都认为它像惰性气体，但更新近的实验支持该元素的行为类似金属铅的观点，这和按它在第14族的位置预测的结果是一样的。这个结论似乎可以说明，化学周期性是一个相当稳固的现象。即使高速电子运动导致的强大的相对论效应，似乎都不能推翻这项150年前得出的简单科学发现。

122
123

第十章

周期表的形式

前面的九章已经说了很多周期表的故事，但有一个很重要的方面还没有提到。这个问题是：为什么有这么多周期表出版，为什么现在课本、论文和网上有这么多周期表？是否存在一份“最佳”周期表？这个问题是否有意义？如果有，我们在发现这样一份最佳周期表的过程中取得了什么进展？

爱德华·马楚斯在关于周期表历史的经典著作中，收录了自19世纪60年代人们第一次整理出周期表以来大约700种周期表的参考资料和图片。而马楚斯的书出版至今又过了四五十年，其间至少又出现了300种表，如果算上网上发布的新周期系统就更多了。存在这么多周期表是需要解释的。当然，其中很多并没有带来什么新东西，还有一些从科学的角度看甚至都自相矛盾。但即便我们去掉这些有问题的表，留下的仍然数量可观。

在第一章中，我们看到周期表有三种基本类型，短式、中长式和长式。尽管这三类大体都传达了类似的信息，但我们之前 124

也说明过，它们处理同价元素分族的方式不同。另外，还有的周期表看上去不像字面意义的表格那样四四方方。其中一类重要的变体是圆形和椭圆形周期系统，它们在强调元素连续性方面或许比方形表更好。方形表中一个周期之末与下一个周期之始的两个元素，如氖和钠、氩和钾之间是断开的，在圆形或椭圆形系统中是连续的。但周期的长度是变化的，和钟面上那种均匀周期不同。于是，圆形周期表的设计者要顾及包括过渡元素在内的几个长周期。例如，本费表（图37）的做法就是用鼓包放置过渡元素，从圆形系统的主体中突出一块。此外还有三维周期表，比如加拿大蒙特利尔的费尔南多·杜富尔设计的这份（图38）。

但接下来我要说的是，这些变体只涉及周期系统的形状变化，它们之间并没有根本区别。构成重大变化的是，将一个或多个元素放到它们在常规周期表中通常位置之外的其他族里。但在开始讨论这一点之前，我们先缓一缓，来大体谈一谈周期表的设计。

周期表好像是一个简单概念，至少看上去如此。正是这点吸引了业余科学家们一试身手，提出一些新版周期表，还总说它们有些性质比之前出版的系统都要好。在一些情况下，业余者，或者说化学、物理的圈外人确实做出了很大的贡献。例如，第六 125 章中提到的安东·范登布鲁克就是经济学家，他最先想到了原子序数的概念，并在《自然》等期刊上发表了数篇有关这一概念的论文。还有一个例子是法国工程师夏尔·雅内，他在1929年发表了已知的第一版左阶梯周期系统，从此持续吸引着周期表 127 的专家和业余爱好者的关注（图39）。

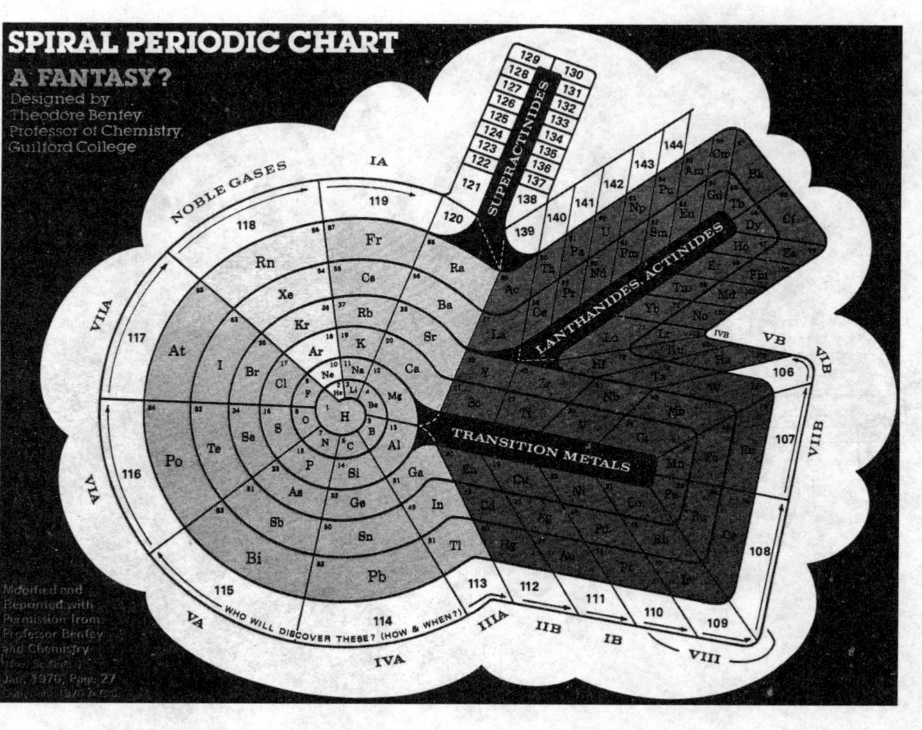
SPIRAL PERIODIC CHART
A FANTASY?
Designed by
Theodore Benfey
Professor of Chemistry
Guilford College
NOBLE GASES
IA
VIIA
VIA
VA
IVA
IIIA
IIB
IB
VIII
VIIB
VIB
VB
IVB
SUPERACTINIDES
LANTHANIDES, ACTINIDES
TRANSITION METALS
WHO WILL DISCOVER THESE? (HOW & WHEN?)
H
He
Li
Be
B
C
N
O
F
Ne
Na
Mg
Al
Si
P
S
Cl
Ar
K
Ca
Ga
Ge
As
Se
Br
Kr
Rb
Sr
In
Sn
Sb
Te
I
Xe
Cs
Ba
Tl
Pb
Bi
Po
At
Rn
Fr
Ra
La
Ac
Ce
Th
Pr
Pa
Nd
U
Pm
Np
Sm
Pu
Eu
Am
Gd
Cm
Tb
Bk
Dy
Cf
Ho
Er
Fm
Tm
Md
Yb
No
Lu
Lr
Hf
Ta
W
Re
Os
Ir
Pt
Au
Hg
Sc
Ti
V
Cr
Mn
Fe
Co
Ni
Cu
Zn
Y
Zr
Nb
Mo
Tc
Ru
Rh
Pd
Ag
Cd
106
107
108
109
110
111
112
113
114
115
116
117
118
119
120
121
122
123
124
125
126
127
128
129
130
131
132
133
134
135
136
137
138
139
140
141
142
143
144
Modified and
Reprinted with
Permission from
Professor Benfey
and Chemistry
Jan. 1970, Page 27

图37　本费的周期系统

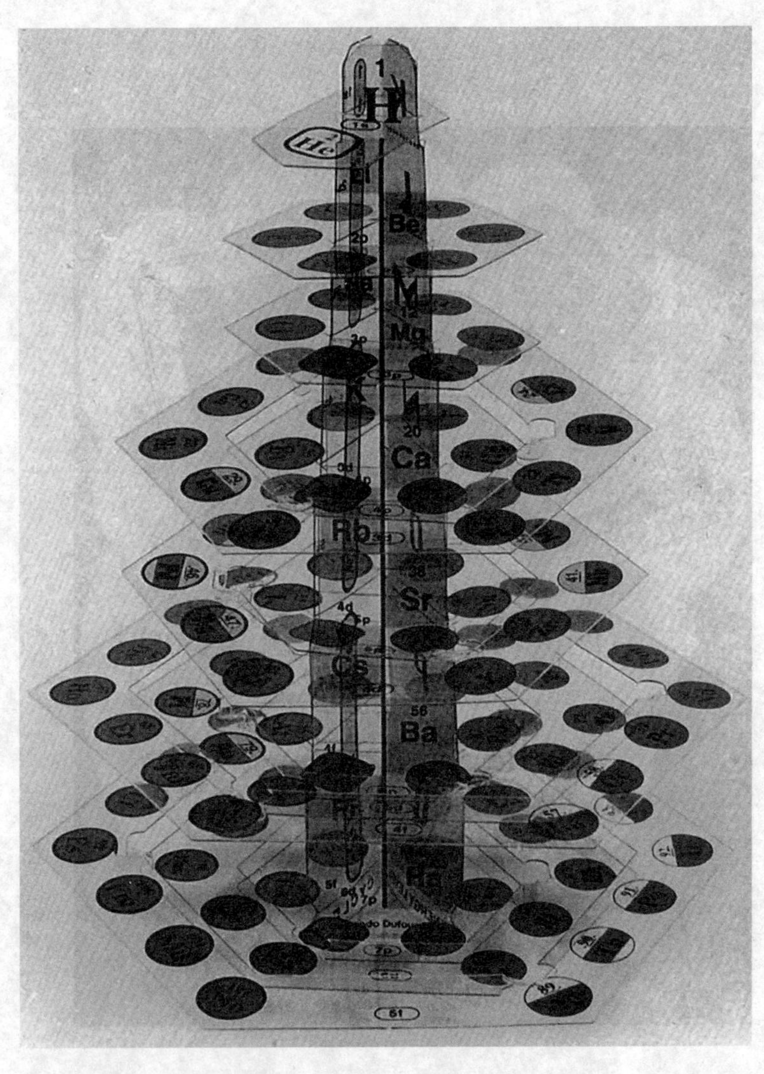

图38　杜富尔的周期树

																														H	He
																														Li	Be
																								B	C	N	O	F	Ne	Na	Mg
																								Al	Si	P	S	Cl	Ar	K	Ca
														Sc	Ti	V	Cr	Mn	Fe	Co	Ni	Cu	Zn	Ga	Ge	As	Se	Br	Kr	Rb	Sr
														Y	Zr	Nb	Mo	Tc	Ru	Rh	Pd	Ag	Cd	In	Sn	Sb	Te	I	Xe	Cs	Ba
La	Ce	Pr	Nd	Pm	Sm	Eu	Gd	Tb	Dy	Ho	Er	Tm	Yb	Lu	Hf	Ta	W	Re	Os	Ir	Pt	Au	Hg	Tl	Pb	Bi	Po	At	Rn	Fr	Ra
Ac	Th	Pa	U	Np	Pu	Am	Cm	Bk	Cf	Es	Fm	Md	No	Lr	Rf	Db	Sg	Bh	Hs	Mt	Ds	Rg	Cn	Nh	Fl	Mc	Lv	Ts	Og	119	120

图39　夏尔·雅内的左阶梯周期表

现在，想想我们之前提的另一个问题是什么？寻找最佳周期表是否有意义？或者说，业余爱好者连同在这个问题上花费时间的专家是否在自欺欺人？我认为，问题的答案藏在每个人对周期系统的哲学观念背后。一方面，如果一个人认为元素性质近似重复的现象是自然世界的一个客观事实，那么他采取的就是唯实论者的态度。对他来说，寻找最佳周期表的问题确实很有意义。最佳周期表就能最好地呈现物质在化学周期性方面的事实，尽管这种表目前还没编制出来。

另一面，对元素周期表持工具主义或者反现实主义观点的人大概会认为，元素的周期性是人类强加给自然的性质。如果确实如此，就不必满腔热情地去寻找**最佳**周期表了，因为这种东西根本不存在。对这样的传统主义者或者反现实主义者来说，用多么精确的方式将元素呈现出来并不是什么紧要的事情，因为他们认为，这只是在处理元素间的人工关系，而不是自然关系。

顺便声明一下我的立场。就周期表而言，我是地道的现实主义者。例如，我很惊讶，许多化学家采取反现实主义的姿态看待周期表；如果问他们氢元素究竟属于第1族（碱金属）还是第17族（卤素），有些化学家的反应是这无关紧要。

在我们详细剖析可能存在的最佳周期表和其他周期表之前，还需要提及一些终极的普遍问题。一个问题是，不同周期表的实用性如何？许多科学家倾向于支持这种或那种特定形式的周期表，因为它在天文学家、地质学家、物理学家等学者的科学工作中或许更有用。这些表主要为实用性而设计。还有的表追求突出元素的“真相”（抱歉没有更好的词语来表达），而不是为方便某一领域的科学家使用。无须多言，任何对最佳周期表的要求都应当避免实用性的问题，尤其应避免问它是否适用于某个特定的学科或子学科。而且，寻求元素真相的表，如果它以某种方式成功捕捉到元素的本质和元素之间的关系，那么它首先就很可能转化为对不同学科有用的表。但我们应该把这种实用性视为额外奖励，而不应该由此决定最佳周期表应具有什么样的形式。

还有对称性的重要问题，这也相当棘手。许多另类周期表的拥护者说，他们的表好在元素以更对称、更规范，或多少更优雅、更美观的方式呈现。科学中对称和美的问题已经讨论了很多，但就所有美学问题而言，一个人视某事物为美，另一个人或许并不这么看。还有，必须小心不要将自然实际上没有的美或规则强加给它。有太多另类周期表的拥护者完全只主张表现形式上的规整，有时忘记了自己只是在谈形式，而不是化学世界本身。

一些特殊情况

这些前提都谈完之后，我们可以介绍一些人们提出的新表。当然我也表示过，我们假设寻找最佳周期表是有意义的。我们从左阶梯周期表开始吧，它是那种本质上另类的周期系统之一，许多元素放在不同于常规周期表的族里。左阶梯表由夏尔·雅内于1929年首先提出，当时量子力学刚发展起来不久。不过，雅内编制的表格似乎与量子力学无关，而完全建立在美学基础上。但研究者很快便看出，比起传统表，左阶梯表有几大特征与量子力学对原子的描述关联更紧密。 130

将氦元素从惰性气体（第18族）的首位移到碱土金属（第2族）的首位，将此时表中左边两族元素断开放到右边组成新表，就得到了左阶梯表。此外，镧系和锕系的28种元素在周期表中通常以类似脚注的形式出现，在此新表中则移到左侧。经过移动，这些元素位列过渡金属元素区左边，完全融入了周期表。

新表的一个优点是表的整体形状更规整、更统一。此外，我们现在得到了两个有两种元素的极短周期，而常规周期表中只能看到一个。相比于常规周期表有一个反常的、不再重复的周期，左阶梯表具有这样的特征：所有周期长度都会重复一次，构成2、2、8、8、18、18等的序列。这些优点都与量子力学无关，但它们是雅内欣赏的特点（雅内并不关心量子力学）。我们在第八章中看到，量子力学引入周期表后，我们能够以电子构型为基础理解周期表。按这种方法，周期表中元素之间的差别在于，区分电子占据的轨道不同。

在传统表中，最左侧的两族元素构成s区，因为它们的区分电子进入s轨道。向右移动，我们会到d区，然后是p区，最后是f区。f区藏在周期表主体的下方。从左向右，这些区的顺序并不是最“自然”的或我们预料中的，因为轨道的能量递增顺序是：

$$s<p<d<f$$

左阶梯表保持了这个顺序，尽管是反过来的。不过，这是否真的算优点尚待讨论，因为任何循环表都有这个特征。

不过，从量子力学的观点看，这份表还有另一个优势。氦原子的电子构型是两个电子都在1s轨道，这是没有争议的事实。氦因此是s区元素。但在常规周期表中，氦因为化学性质非常不活泼而被归为惰性气体，就像其他惰性气体（氖、氩、氪、氙、氡）一样。

这个情形似乎和之前讨论过的碲碘调换历史事件相似。那一次，研究者不得不为保留化学相似性而忽略原子量顺序。同样，现在氦的情况也有两种可能：

1. 电子结构不是元素归属哪一族的最终决定因素，在适当的时候就会被某种新的标准取代。（例如，在元素排序上，原子序数最终取代了原子量，由此解决了调换对的问题。）

2. 我们实际上并没有遇到两种势均力敌的情况，电子构型仍然是主导因素，我们由此应该忽略氦明显的化学惰性。

注意，选项1实际上倾向于常规周期表，而选项2倾向于左阶梯表。显然，从量子力学的观点看，我们很难确定左阶梯表是否体现出了优势。现在我再提出另一个想法。回想一下，第四章在谈元素的本质时我们说了什么，尤其是门捷列夫怎样对待元素：他喜欢更抽象的意义，而不是把元素和单质或者说分离出来的物质捆绑起来。对元素抽象意义的依赖，可以为将氦移动到碱土族提供合理解释。考虑到氦化学活动性低，而不将它列入更活泼的碱土元素的做法，也可用上述理由反驳，即我们的关注点应当放在元素的本性上，把它看作抽象的实体，而不应当放在化学性质上。无论如何，这一移动都相当于在说，如果可以忽略氦的化学性质，那"为什么不"把氦列入碱土族呢？

本节最后，我再谈一项2017年做的有趣实验：阿尔捷姆·奥加诺夫带领团队制成了真正的氦钠化合物Na_2He，诚然，它是在极高的压力下制成的。存在这种一目了然的真正的氦化合物，表明惰性气体中最不活泼的是氖而不是氦，这重启了左阶梯周期表能否代表最佳周期表的争论。

将原子序数三元素组应用于第3族

关于周期表的第3族，化学家和化学教育工作者之间仍长期存在分歧。一些旧周期表的第3族为如下元素：

Sc
Y
La
Ac

后来，许多教科书中周期表的第3族开始变成这样：

Sc
Y
Lu
Lr

其论据是推定的电子构型。1986年，辛辛那提大学的威廉·詹森发表论文称，教科书编写者和周期表设计者都应当将第3族定为Sc、Y、Lu和Lr。

再后来，一些学者负隅顽抗，主张回归Sc、Y、La和Ac。如果真的回归，那么将原子序数三元素组的概念用在第3族，会得出什么结果呢？如果我们考虑原子序数三元素组，那么答案依然确定无疑，即倾向于支持詹森的分族。第一个三元素组是准确的，

Y	39
Lu	$71 = (39 + 103)/2$
Lr	103

第二个三元素组不正确，

Y	39
La	$57 \neq (39 + 89)/2 = 64$
Ac	89

134

但詹森的分族之所以更好，还有一个原因，这个原因完全不需要依据原子序数三元素组。

若考虑长式周期表，并把镥、铹或镧、锕放到第3族，那么只有第一种排列说得通，因为那样排列原子序数是连续增大的。相反，把镧、锕放入长式表的第3族，从原子序数递增的角度看，就会出现两种相当刺眼的反常情况（图40）。

最后，我们实际上还有第三种可能，但要将d区元素非常别扭地拆开，如图41所示。尽管一些书收录了类似图41的周期表，但这不是很常见的设计，原因相当明显。用这种方式呈现周期表，需要将周期表的d区分成非常不均匀的两部分，一部分只有一个元素宽，另一部分则有九个元素宽。考虑到这种拆分在周期表的其他区都没有出现过，因此在三份表中，它似乎最不可能反映元素自然的真实排布。

应当承认，在传统周期表中，其实有一个区——s区——的第一行通常被分成两部分，即在传统的18或32列表中，氢和氦

H 1																															He 2
Li 3	Be 4																									B 5	C 6	N 7	O 8	F 9	Ne 10
Na 11	Mg 12																									Al 13	Si 14	P 15	S 16	Cl 17	Ar 18
K 19	Ca 20															Sc 21	Ti 22	V 23	Cr 24	Mn 25	Fe 26	Co 27	Ni 28	Cu 29	Zn 30	Ga 31	Ge 32	As 33	Se 34	Br 35	Kr 36
Rb 37	Sr 38															Y 39	Zr 40	Nb 41	Mo 42	Tc 43	Ru 44	Rh 45	Pd 46	Ag 47	Cd 48	In 49	Sn 50	Sb 51	Te 52	I 53	Xe 54
Cs 55	Ba 56	La 57	Ce 58	Pr 59	Nd 60	Pm 61	Sm 62	Eu 63	Gd 64	Tb 65	Dy 66	Ho 67	Er 68	Tm 69	Yb 70	*Lu* 71	Hf 72	Ta 73	W 74	Re 75	Os 76	Ir 77	Pt 78	Au 79	Hg 80	Tl 81	Pb 82	Bi 83	Po 84	At 85	Rn 86
Fr 87	Ra 88	Ac 89	Th 90	Pa 91	U 92	Np 93	Pu 94	Am 95	Cm 96	Bk 97	Cf 98	Es 99	Fm 100	Md 101	No 102	*Lr* 103	Rf 104	Db 105	Sg 106	Bh 107	Hs 108	Mt 109	Ds 110	Rg 111	Cn 112	Nh 113	Fl 114	Mc 115	Lv 116	Ts 117	Og 118

H 1																															He 2
Li 3	Be 4																									B 5	C 6	N 7	O 8	F 9	Ne 10
Na 11	Mg 12																									Al 13	Si 14	P 15	S 16	Cl 17	Ar 18
K 19	Ca 20															Sc 21	Ti 22	V 23	Cr 24	Mn 25	Fe 26	Co 27	Ni 28	Cu 29	Zn 30	Ga 31	Ge 32	As 33	Se 34	Br 35	Kr 36
Rb 37	Sr 38															Y 39	Zr 40	Nb 41	Mo 42	Tc 43	Ru 44	Rh 45	Pd 46	Ag 47	Cd 48	In 49	Sn 50	Sb 51	Te 52	I 53	Xe 54
Cs 55	Ba 56	Ce 58	Pr 59	Nd 60	Pm 61	Sm 62	Eu 63	Gd 64	Tb 65	Dy 66	Ho 67	Er 68	Tm 69	Yb 70	Lu 71	*La* 57	Hf 72	Ta 73	W 74	Re 75	Os 76	Ir 77	Pt 78	Au 79	Hg 80	Tl 81	Pb 82	Bi 83	Po 84	At 85	Rn 86
Fr 87	Ra 88	Th 90	Pa 91	U 92	Np 93	Pu 94	Am 95	Cm 96	Bk 97	Cf 98	Es 99	Fm 100	Md 101	No 102	Lr 103	*Ac* 89	Rf 104	Db 105	Sg 106	Bh 107	Hs 108	Mt 109	Ds 110	Rg 111	Cn 112	Nh 113	Fl 114	Mc 115	Lv 116	Ts 117	Og 118

图40　表现镥和铹的两种分布的长式周期表。只有上方的版本能确保列内的原子序数是连续的

H																															He
Li	Be																									B	C	N	O	F	Ne
Na	Mg																									Al	Si	P	S	Cl	Ar
K	Ca	Sc															Ti	V	Cr	Mn	Fe	Co	Ni	Cu	Zn	Ga	Ge	As	Se	Br	Kr
Rb	Sr	Y															Zr	Nb	Mo	Tc	Ru	Rh	Pd	Ag	Cd	In	Sn	Sb	Te	I	Xe
Cs	Ba	La	Ce	Pr	Nd	Pm	Sm	Eu	Gd	Tb	Dy	Ho	Er	Tm	Yb	Lu	Hf	Ta	W	Re	Os	Ir	Pt	Au	Hg	Tl	Pb	Bi	Po	At	Rn
Fr	Ra	Ac	Th	Pa	U	Np	Pu	Am	Cm	Bk	Cf	Es	Fm	Md	No	Lr	Rf	Db	Sg	Bh	Hs	Mt	Ds	Rg	Cn	Nh	Fl	Mc	Lv	Ts	Og

图41　第三种呈现长式周期表的方式，其中d区分成了不均匀的两部分，各有一个族和九个族

是分开的。不过，这一分离以更合理的对称形式呈现，此外我们
135 看到，雅内的左阶梯表完全回避了这种情况。

氢的情况暨重新考虑原子序数三元素组

传统上引发问题的另一个元素正是第1号元素氢。一方面，从化学上讲，由于氢能够形成+1离子H^+，它似乎属于第1族（碱金属）。但另一方面，氢非常特殊，因为它还可以形成 −1离子H^-，比如在NaH和CaH_2等金属氢化物中。这一行为支持将氢放在第17族（卤素），它们也形成 −1离子。这个难题如何解决呢？一些学者采取了轻巧的方式，让氢尊贵地“漂浮”在周期表主体上方。换言之，他们从两个可能的位置中选定一个。

这在我眼里就是“化学精英主义”的表现，它似乎在说，虽然所有元素都服从周期律，但不知为何，氢是个特例，凌驾于法则之上，像极了过去的英国王室。从本书第一版出版到最近，我都主张构建新的原子序数三元素组，以解决氢的位置归属问题。如果你用这种方法，就会得到一个明显结果，倾向于将氢放在卤素，而不是碱金属中。常规周期表将氢放在碱金属里，构不成完美的三元素组，而如果让氢居于卤素的顶端，就能得到一个新的原子序数三元素组（图42）：

H	1		H	1	
Li	3	$(1+11)/2 \neq 3$	F	9	$(1+17)/2 = 9$
Na	11		Cl	17	

尽管这个建议看起来非常吸引人，但我现在认为，这可能是
138 一项错误的策略。我这么说的理由是，各族元素的第一个成员

																												H	He	Li	Be
																								B	C	N	O	F	Ne	Na	Mg
																								Al	Si	P	S	Cl	Ar	K	Ca
														Sc	Ti	V	Cr	Mn	Fe	Co	Ni	Cu	Zn	Ga	Ge	As	Se	Br	Kr	Rb	Sr
														Y	Zr	Nb	Mo	Tc	Ru	Rh	Pd	Ag	Cd	In	Sn	Sb	Te	I	Xe	Cs	Ba
La	Ce	Pr	Nd	Pm	Sm	Eu	Gd	Tb	Dy	Ho	Er	Tm	Yb	Lu	Hf	Ta	W	Re	Os	Ir	Pt	Au	Hg	Tl	Pb	Bi	Po	At	Rn	Fr	Ra
Ac	Th	Pa	U	Np	Pu	Am	Cm	Bk	Cf	Es	Fm	Md	No	Lr	Rf	Db	Sg	Bh	Hs	Mt	Ds	Rg	Cn	Nh	Fl	Mc	Lv	Ts	Og	119	120

图42　以使原子序数三元素组数量最多为原则编制的周期表

都不是三元素组的成员，没有理由认为如卤素这一族例外。

研究正确的三元素组在周期表中出现的条件是一件很有意思的事。在以s区元素位于周期表左侧为特点的常规周期表中，可以看到一个反常现象。单就s区元素来说，只有第一、第二个元素所在的周期等长时，原子序数三元素组才会出现，如锂（3）、钠（11）和钾（19）。然而，在p区、d区，原则上甚至还有f区，只有在第二、第三个元素所在的周期等长时，三元素组才会出现，如氯（17）、溴（35）和碘（53）。

但是，如果将元素用左阶梯表的形式呈现，那么所有三元素组都由第二、第三个元素所在的周期等长的三个元素组成，无一例外（图43）。我认为，这或许是支持左阶梯周期表优越性的又一个论据，尽管它只是形式上的。我还认为，这份表或许就是人们一直在寻找的最佳周期表。

国际纯粹与应用化学联合会是否有官方的周期表？

国际纯粹与应用化学联合会是化学命名的管理机构，对最佳形式的周期表持有相当模糊的立场。尽管国际纯粹与应用化学联合会的官方政策不会推荐任何特定的形式，但他们的文献中经常出现如图44所示的周期表，其中包括15个元素宽度的镧系和锕系。这种形式的缺点是，它使表中的第3族只剩下两种元素，而其他任何族都没有出现这种情况。将这种周期表用如图45所示的32列表呈现，反常就更明显了。我认为，现在是时候了，国际纯粹与应用化学联合会应该认识到这种形式的问题，并应该对现有最一致的周期表做出正式裁决，如我之前建议的一样，收录图40上方的版本。

																														1	2
																														3	4
																								5	6	7	8	9	10	11	12
																								13	14	15	16	17	18	19	20
														21	22	23	24	25	26	27	28	29	30	31	32	33	34	35	36	37	38
														39	40	41	42	43	44	45	46	47	48	49	50	51	52	53	54	55	56
57	58	59	60	61	62	63	64	65	66	67	68	69	70	71	72	73	74	75	76	77	78	79	80	81	82	83	84	85	86	87	88
89	90	91	92	93	94	95	96	97	98	99	100	101	102	103	104	105	106	107	108	109	110	111	112	113	114	115	116	117	118	119	120

图43　在左阶梯周期表中高亮显示的原子序数三元素组。每个三元素组的第二、第三个元素所在的周期都等长

族 #

1	2	3	4	5	6	7	8	9	10	11	12	13	14	15	16	17	18
H																	He
Li	Be											B	C	N	O	F	Ne
Na	Mg											Al	Si	P	S	Cl	Ar
K	Ca	Sc	Ti	V	Cr	Mn	Fe	Co	Ni	Cu	Zn	Ga	Ge	As	Se	Br	Kr
Rb	Sr	Y	Zr	Nb	Mo	Tc	Ru	Rh	Pd	Ag	Cd	In	Sn	Sb	Te	I	Xe
Cs	Ba		Hf	Ta	W	Re	Os	Ir	Pt	Au	Hg	Tl	Pb	Bi	Po	At	Rn
Fr	Ra		Rf	Db	Sg	Bh	Hs	Mt	Ds	Rg	Cn	Nh	Fl	Mc	Lv	Ts	Og

La	Ce	Pr	Nd	Pm	Sm	Eu	Gd	Tb	Dy	Ho	Er	Tm	Yb	Lu
Ac	Th	Pa	U	Np	Pu	Am	Cm	Bk	Cf	Es	Fm	Md	No	Lr

图44　国际纯粹与应用化学联合会的周期表，其中镧系和锕系元素区为15个元素宽度。第3族只包括两个元素

我希望读者在读完本书之后认识到，尽管周期表被构想出来已经过了150年了，但它依旧是一个充满趣味、不断发展的事物。

H																																He
Li	Be																										B	C	N	O	F	Ne
Na	Mg																										Al	Si	P	S	Cl	Ar
K	Ca																Sc	Ti	V	Cr	Mn	Fe	Co	Ni	Cu	Zn	Ga	Ge	As	Se	Br	Kr
Rb	Sr																Y	Zr	Nb	Mo	Tc	Ru	Rh	Pd	Ag	Cd	In	Sn	Sb	Te	I	Xe
Cs	Ba	La	Ce	Pr	Nd	Pm	Sm	Eu	Gd	Tb	Dy	Ho	Er	Tm	Yb	Lu		Hf	Ta	W	Re	Os	Ir	Pt	Au	Hg	Tl	Pb	Bi	Po	At	Rn
Fr	Ra	Ac	Th	Pa	U	Np	Pu	Am	Cm	Bk	Cf	Es	Fm	Md	No	Lr		Rf	Db	Sg	Bh	Hs	Mt	Ds	Rg	Cn	Nh	Fl	Mc	Lv	Ts	Og

图45　同国际纯粹与应用化学联合会版直接对应的32列周期表，其中镧系和锕系为15个元素宽度

索 引

(条目后的数字为原书页码，见本书边码)

C

索引

D

E

I

J

K

L

M

N

O

P

Q

R

S

T

U

V

W

X

Y

Z

Eric R. Scerri

THE PERIODIC TABLE

A Very Short Introduction

SECOND EDITION

Contents

Acknowledgements and Dedication

I acknowledge the help of all editors and staff at OUP, including Jeremy Lewis for suggesting the idea of a *Very Short Introduction*. I also thank colleagues, students, and librarians who helped me during the preparation of this book.

This book is dedicated to my wife Elisa Seidner.

Preface

Much has been written about the wonders of the periodic table. Here are just a few examples.

> The Periodic Table is nature's Rosetta stone. To the uninitiated, it's just 100-plus numbered boxes, each containing one or two letters, arranged with an odd, skewed symmetry. To chemists, however, the periodic table reveals the organizing principles of matter, which is to say, the organizing principles of chemistry. At a fundamental level, all of chemistry is contained in the periodic table.
>
> That's not to say, of course, that all of chemistry is obvious from the periodic table. Far from it. But the structure of the table reflects the electronic structure of the elements, and hence their chemical properties and behavior. Perhaps it would be more appropriate to say that all of chemistry starts with the periodic table.
>
> (Rudy Baum, *C&EN Special Issue on Elements*)

Astronomer Harlow Shapley writes:

> The periodic table is probably the most compact and meaningful compilation of knowledge that man has yet devised. The periodic table does for matter what the geological age table does for cosmic time. Its history is the story of man's great conquests in the microcosmos.

As Robert Hicks, a historian of chemistry, says on an Internet podcast:

> Perhaps the most recognizable icon in all of science is the periodic table of elements. This chart has become our model for how atoms and molecules arrange themselves to create matter as we know it. How the world is organized on the most minute level. Throughout history the periodic table has changed. Newly discovered elements have been added to it and other elements have been disproved and either modified or removed. In this way the periodic table acts like a store of the history of chemistry, a template for current developments and a basis for the future of the chemical sciences... a map of the world's most basic building blocks.

As a final example for now, here is C. P. Snow, a physical chemist known for his writings on the 'two cultures':

> [On learning about the table for] the first time I saw a medley of haphazard facts fall into line and order. All the jumbles and recipes and hotchpotch of the inorganic chemistry of my boyhood seemed to fit themselves into the scheme before my eyes—as though one were standing beside a jungle and it suddenly transformed itself into a Dutch garden.

What is remarkable about the periodic table is its simultaneous simplicity and familiarity coupled with its truly fundamental status in science. The simplicity is alluded to in the above quotations. The periodic table seems to organize the fundamental components of all matter. It is also familiar to most people. Almost everybody with even just an elementary acquaintance with chemistry can recall the existence of the periodic table even after everything else they ever learned in chemistry may have been forgotten. The periodic table is almost as familiar as the chemical formula for water. It has become a true cultural icon that is used by artists, advertisers, and of course scientists of all kinds.

At the same time, the periodic table is more than just a device for teaching and learning chemistry. It reflects the natural order of things in the world and, as far as we know, in the whole universe. It consists of groupings of elements into vertical columns. If a chemist or even a student of chemistry knows the properties of one typical element in any group, such as sodium, he or she will have a good idea of the properties of elements in the same group, such as potassium, rubidium, and caesium.

More fundamentally, the order inherent in the periodic table has led to a deep knowledge of the structure of the atom and the notion that electrons essentially encircle the nucleus in specific shells and orbitals. These arrangements of electrons in turn serve to rationalize the periodic table. They explain why, broadly speaking, the elements sodium, potassium, rubidium, and so on fall into the same group in the first place. But more importantly, the understanding of atomic structure that was first arrived at in trying to understand the periodic table has been applied in many other fields of science. This knowledge has contributed to the development of, first, the old quantum theory and then to its more mature cousin, quantum mechanics, a body of knowledge that continues to be the fundamental theory of physics that can explain the behaviour not just of all matter, but also all forms of radiation such as visible light, X-rays, and ultraviolet radiation.

Unlike most scientific discoveries made in the 19th century, the periodic table has not been refuted by discoveries made in the 20th and 21st centuries. Rather, discoveries in modern physics, in particular, have served to refine the periodic table and to tidy up some remaining anomalies. But its overall form and validity have remained intact as another testament to the power and depth of this system of knowledge.

Before examining the periodic table, we will consider its occupants, the elements. Then we will take a quick look at the modern periodic table and some of its variants, before looking into its history and how we got to our present level of understanding, from Chapter 3.

Preface to the second edition

The people in charge of the *Very Short Introduction* series have asked me to produce a second edition to coincide with the 150th anniversary of Mendeleev's discovery of the mature periodic table and the fact that 2019 has been proclaimed as the International Year of the Periodic Table (IYPT) by UNESCO.

Needless to say, I welcome this opportunity to correct any errors in the first edition that have been pointed out by readers and to bring matters up to date, since a number of developments have occurred in the study of the periodic table and related issues.

For example, the important 4s–3d orbital occupation question concerning the modern account of the periodic table now has a satisfactory explanation that fully explains the relative order of filling and ionization of these atomic orbitals. In addition, the relevance of the Aufbau principle and anomalies to it has been reconceptualized and will be briefly explained in this updated version.

There have been new developments in the field of the synthetic elements. For example, the ratification and naming of elements 113, 115, 117, and 118 as nihonium, moscovium, tennessine, and oganesson has taken place since the first edition appeared. This development also represents the completion of the seventh period for the first time since the periodic table was discovered. Nevertheless, work has begun on trying to synthesize elements in the next period, starting with elements 119 and 120. The g-block elements will begin, at least formally, just beyond that point at element 121.

There have been some new developments concerning the question of which elements should occupy group 3 of the periodic table as well as new experiments that have a bearing on the question of the placement of helium.

I hope the reader might be encouraged to contact me with any questions or suggestions on any issues that have been raised in this book.

Eric Scerri
Los Angeles
2019

List of illustrations

Chapter 1
The elements

The ancient Greek philosophers recognized just four elements—earth, water, air, and fire—all of which survive in the astrological subdivision of the twelve signs of the zodiac. Some of those philosophers believed that these different elements consisted of microscopic components with differing shapes, and that this explained the various properties of the elements. The basic shapes of the four elements were thought to be those of the Platonic solids (Figure 1) made up entirely of the same two-dimensional shapes like triangles or squares. The Greeks believed that earth consisted of microscopic cubic particles. This association was made because, of all the Platonic solids, the cube possesses the faces with the largest surface area. The liquidity of water was explained by an appeal to the smoother shape possessed by the icosahedron, while fire was said to be painful to the touch because it consisted of the sharp particles in the form of tetrahedra. Air was thought to consist of octahedra since that was the only remaining Platonic solid. Sometime later, a fifth Platonic solid, the dodecahedron, was discovered by mathematicians, and this led Aristotle to propose that there might be a fifth element, or 'quintessence', which also became known as ether.

Today, the notion that elements are made up of Platonic solids is regarded as incorrect, but it was the origin of the fruitful

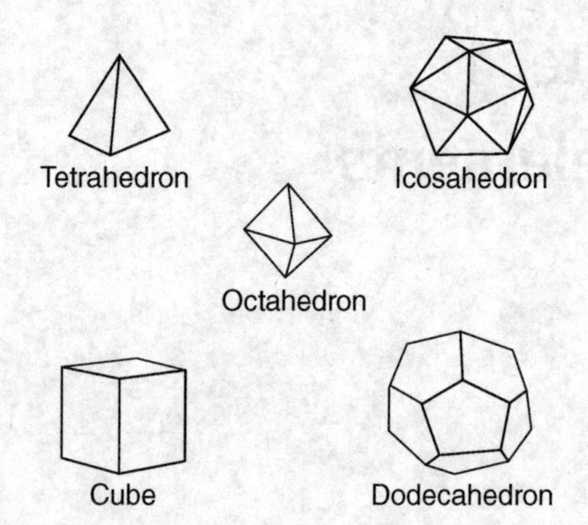

1. The Platonic solids, each associated with one of the ancient elements.

notion that macroscopic properties of substances are governed by the structures of the microscopic components of which they are comprised. These 'elements' survived well into the Middle Ages and beyond, augmented with a few others discovered by the alchemists, the precursors of modern-day chemists. The best-known goal of the alchemists was to bring about the transmutation of elements. In particular, they attempted to change the base metal lead into the noble metal gold, whose colour, rarity, and chemical inertness has made it one of the most treasured substances since the dawn of civilization.

But in addition to being regarded as substances that could actually exist, Greek philosophers thought of the 'elements' as principles, or as tendencies and potentialities that gave rise to the observable properties of the elements. This rather subtle distinction between the abstract form of an element and its observable form has played an important role in the development of chemistry, although these days the subtler meaning is not very well understood even by professional chemists. The notion of an abstract element has nonetheless served as a fundamental guiding principle to some of the pioneers of the periodic system such as Dmitri Mendeleev, its major discoverer.

According to most textbook accounts, chemistry began properly only when it turned its back on ancient Greek wisdom and alchemy and on this seemingly mystical understanding of the nature of elements. The triumph of modern science is generally regarded as resting on direct experimentation, which holds that only what can be observed should count. Not surprisingly, the subtler and perhaps more fundamental sense of the concept of elements has generally been rejected. For example, Antoine Lavoisier took the view that an element should be defined by an appeal to empirical observation, thus relegating the role of abstract elements or elements as principles. Lavoisier held that an element should be defined as a material substance that has yet to be broken down into any more fundamental components. In 1789, Lavoisier published a list of thirty-three simple substances, or elements, according to this empirical criterion (Figure 2). The ancient elements of earth, water, air, and fire, which had by now been shown to consist of simpler substances, were correctly omitted from the list of elements.

Many of the substances in Lavoisier's list would qualify as elements by modern standards, while others like *lumière* (light) and *calorique* (heat) are no longer regarded as elements. Rapid advances in techniques of separation and characterization of chemical substances over the forthcoming years would help chemists expand this list. The important technique of spectroscopy, which measures the emission and absorption spectra of various kinds of radiation, would eventually yield a very accurate means by which each element could be identified through its 'fingerprint'. Today, we recognize about ninety naturally occurring elements. Moreover, an additional twenty-five or so elements have been artificially synthesized.

	Noms nouveaux.	*Noms anciens correſpondans.*
Subſtances ſimples qui appartiennent aux trois règnes, & qu'on peut regarder comme les élémens des corps.	Lumière	Lumière.
	Calorique	Chaleur. Principe de la chaleur. Fluide igné. Feu. Matière du feu & de la chaleur.
	Oxygène	Air déphlogiſtiqué. Air empiréal. Air vital. Baſe de l'air vital.
	Azote	Gaz phlogiſtiqué. Mofète. Baſe de la mofète.
	Hydrogène	Gaz inflammable. Baſe du gaz inflammable.
Subſtances ſimples non métalliques oxidables & acidifiables.	Soufre	Soufre.
	Phoſphore	Phoſphore.
	Carbone	Charbon pur.
	Radical muriatique	Inconnu.
	Radical fluorique	Inconnu.
	Radical boracique	Inconnu.
Subſtances ſimples métalliques oxidables & acidifiables.	Antimoine	Antimoine.
	Argent	Argent.
	Arſenic	Arſenic.
	Biſmuth	Biſmuth.
	Cobalt	Cobalt.
	Cuivre	Cuivre.
	Etain	Etain.
	Fer	Fer.
	Manganèſe	Manganèſe.
	Mercure	Mercure.
	Molybdène	Molybdène.
	Nickel	Nickel.
	Or	Or.
	Platine	Platine.
	Plomb	Plomb.
	Tungſtène	Tungſtène.
	Zinc	Zinc.
Subſtances ſimples ſalifiables terreuſes.	Chaux	Terre calcaire, chaux.
	Magnéſie	Magnéſie, baſe du ſel d'epſom.
	Baryte	Barote, terre peſante.
	Alumine	Argile, terre de l'alun, baſe de l'alun.
	Silice	Terre ſiliceuſe, terre vitrifiable.

2. Lavoisier's list of elements as simple substances.

The discovery of the elements

Some elements like iron, copper, gold, and silver have been known since the dawn of civilization. This reflects the fact that these elements can occur in uncombined form or are easy to separate from the minerals in which they occur.

Historians and archaeologists refer to certain epochs in human history as the Iron Age or the Bronze Age (bronze is an alloy of copper and tin). The alchemists added several more elements to the list, including sulphur, mercury, and phosphorus. In relatively modern times, the discovery of electricity enabled chemists to isolate many of the more reactive elements, which, unlike copper and iron, could not be obtained by heating their ores with charcoal (carbon). There have been a number of major episodes in the history of chemistry when half a dozen or so elements were discovered within a period of a few years. For example, the English chemist Humphry Davy made use of electricity, or more specifically the technique of electrolysis, to isolate about ten elements, including calcium, barium, magnesium, sodium, and chlorine.

Following the discovery of radioactivity and nuclear fission, yet more elements were discovered. The last seven elements to be isolated within the limits of the naturally occurring elements were protactinium, hafnium, rhenium, technetium, francium, astatine, and promethium, between the years 1917 and 1945. One of the last gaps to be filled was that corresponding to element 43, which became known as technetium from the Greek *techne*, meaning 'artificial'. It was 'manufactured' in the course of radio-chemical reactions that would not have been feasible before the advent of nuclear physics. Nevertheless, it now appears that technetium does occur naturally on earth, although in minuscule amounts.

Names of the elements

Part of the appeal of the periodic table derives from the individual nature of the elements such as their colours or how they feel to the touch. Much interest also lies in their names. The chemist and concentration camp survivor Primo Levi wrote a much-acclaimed book called simply *The Periodic Table* in which each chapter is named after an element. The book is mostly about his relations and acquaintances, but each anecdote is motivated by Levi's love of a particular element. The neurologist and author Oliver Sacks wrote a book called *Uncle Tungsten* in which he tells of his fascination with the elements, with chemistry, and in particular with the periodic table. More recently, two popular books on the elements have been written by Sam Kean and Hugh Aldersey-Williams. I think it is fair to say that the appeal of the elements in the public imagination has now truly arrived.

During the many centuries over which the elements have been discovered, many different approaches have been used to give them their names. Promethium, element 61, takes its name from Prometheus, the god who stole fire from heaven and gave it to human beings only to be punished for this act by Zeus. The connection of this tale to element 61 is in the heroic effort that was needed to isolate it, by analogy to the heroic and dangerous feat of Prometheus in Greek mythology. Promethium is one of the few elements that do not occur naturally on the earth. It was obtained as a decay product from the fission of another element, uranium.

Planets and other celestial bodies have also been used to name some elements. Helium is named after *helios*, the Greek name for the Sun. It was first observed in the spectrum of the Sun in 1868, and it was not until 1895 that it was first identified in terrestrial samples. Similarly, we have palladium after the asteroid Pallas, which in turn was named after Pallas, the Greek goddess of wisdom. The element cerium is named after Ceres, the first asteroid to be discovered, in the year 1801. Uranium is named after the planet Uranus, both the planet and the element having been discovered in the 1780s. In many of these cases too, the mythological theme persists. Uranus, for example, was the god of heaven in Greek mythology.

Many elements get their names from colours. The yellow-green gas chlorine comes from the Greek word *khloros*, which denotes the colour yellow-green. Caesium is named after the Latin *cesium*, which means grey-blue, because it has prominent grey-blue lines in its spectrum. The salts of the element rhodium often have a pink colour, and this explains why the name of the element was chosen from *rhodon*, the Greek for rose. The metal thallium gets its name from the Latin *thallus*, meaning green twig. It is an element that was discovered by the British chemist William Crookes from the prominent green line in its spectrum.

A large number of element names have come from the place where their discoverer lived, or wished to honour, such as americium, berkelium, californium, darmstadtium, europium, francium, germanium, hassium, polonium, gallium, hafnium (from *Hafnia*, the Latin name for Copenhagen), lutetium (from *Lutetia*, Latin for Paris), moscovium, nihonium (from Nihon or Japan), rhenium (from the region of the Rhine river), ruthenium (from *Rus*, Latin for an area which includes present-day western Russia, Ukraine, Belarus, and parts of Slovakia and Poland), and tennessine.
Yet other element names are derived from geographical locations connected with the minerals in which they were found. This category includes the case of four elements named after the Swedish village of Ytterby, which lies close to Stockholm. Erbium, terbium, ytterbium, and yttrium were all found in ores located around this village, while a fifth element, holmium, was named after the Latin name for Stockholm.

In cases of more recently synthesized elements, their names come from those of the discoverer or a person whom the discoverers wished to honour. For example, we have bohrium, curium, einsteinium, fermium, flerovium, lawrencium, meintnerium, mendeleevium, nobelium, oganesson, roentgenium, rutherfordium, and seaborgium.

The naming of the later transuranium elements has featured nationalistic controversies and, in some cases, bitter disputes over who first synthesized the element and who should therefore be given the honour of selecting a name for it. In an attempt to resolve such disputes, the International Union of Pure and Applied Chemistry (IUPAC) decreed that the elements should be named impartially and systematically with the Latin numerals for the atomic number of the element in each case. Element 105, for example, was known as un-nil-pentium (Unp), while element 106 was un-nil-hexium (Unh). But more recently, after much deliberation on some of these later superheavy elements, IUPAC has returned the naming rights to the discoverers or synthesizers who were judged to have established priority in each case. Elements 105 and 106 are now called dubnium and seaborgium respectively.

The symbols that are used to depict each element in the periodic table also have a rich and interesting story. In alchemical times, the symbols for the elements often coincided with those of the planets from which they were named, or with which they were associated (Figure 3). The element mercury, for example, shared the same symbol as that of Mercury, the innermost planet in the solar system. Copper was associated with the planet Venus, and both the element and the planet shared the same symbol.

When John Dalton published his atomic theory in 1805, he retained several of the alchemical symbols for the elements. These were rather cumbersome, however, and did not lend themselves easily to reproduction in articles and books. The modern use of letter symbols was introduced by the Swedish chemist Jöns Jacob Berzelius in 1813.

Metal							
	gold	silver	iron	mercury	tin	copper	lead
Symbol							
	○	☾	♂	☿	♃	♀	♄
Celestial Body							
	Sun	Moon	Mars	Mercury	Jupiter	Venus	Saturn
Day							
Lat.	*Solis*	*Lunae*	*Martis*	*Mercurii*	*Jovis*	*Veneris*	*Saturni*
Fr.	*dimanche*	*lundi*	*mardi*	*mercredi*	*jeudi*	*vendredi*	*samedi*
Eng.	Sunday	Monday	Tuesday	Wednesday	Thursday	Friday	Saturday

3. Names and symbols of ancient elements.

A small minority of elements in the modern periodic table are represented by a single letter of the alphabet. These include hydrogen, carbon, oxygen, nitrogen, sulphur, and fluorine, which appear as H, C, O, N, S, and F. Most elements are depicted by two letters, the first of which is a capital letter and the second a lower-case letter. For example, we have Kr, Mg, Ne, Ba, Sc, for krypton, magnesium, neon, barium, and scandium, respectively. Some of the two-letter symbols are by no means intuitively obvious, such as Cu, Na, Fe, Pb, Hg, Ag, and Au, which are derived from the Latin names for the elements copper, sodium, iron, lead, mercury, silver, and gold. Tungsten is represented by a W after the German name for the element, which is wolfram.

Chapter 2

A quick overview of the modern periodic table

The modern periodic table

The manner in which the elements are arranged in rows and columns in the periodic table reveals many relationships among them. Some of these relationships are very well known while others may still await discovery. In the 1980s, scientists discovered that superconductivity, meaning the flow of an electric current with zero resistance, occurred at much higher temperatures than had previously been observed. From typical values of 20 K or less, the superconducting temperature very quickly leapt to values such as 100 K.

The discovery of these high-temperature superconductors came about when the elements lanthanum, copper, oxygen, and barium were combined together to form a complicated compound. There followed a flurry of worldwide activity in an effort to raise the temperature at which the effect could be maintained. The ultimate goal was to achieve room temperature superconductivity, which would allow technological breakthroughs such as levitating trains gliding effortlessly along superconducting rails. One of the main principles used in this quest was the periodic table of the elements. The table allowed researchers to replace some of the elements in the compound with others that are known to behave in a similar manner, in order to examine the outcome on their superconducting behaviour. This is how the element yttrium was incorporated into a new set of superconducting compounds, to produce a superconducting temperature of 93 K in the compound $YBa_2Cu_3O_7$. This knowledge, and undoubtedly much more, lies dormant within the periodic system waiting to be discovered and put to good use.

Even more recently, a new class of high-temperature superconductors has been discovered. They are the oxypnictides, a class of materials including oxygen, a pnictogen (group 15 elements), and one or more other elements. Interest in these compounds increased dramatically after the publication of the superconducting properties of LaOFeP and LaOFeAs, which were discovered in 2006 and 2008 respectively. Once again, the idea of using arsenic (As), as in the latter compound, came from its position, which lies directly below phosphorus in the periodic table.

Chemical analogies between elements in the same group are also of great interest in the field of medicine. For example, the element beryllium sits at the top of group 2 of the periodic table and above magnesium. Because of the similarity between these two elements, beryllium can replace the element magnesium that is essential to human beings. This behaviour accounts for one of the many ways in which beryllium is toxic to humans, for although it has similar properties, they are not identical. Similarly, the element cadmium lies directly below zinc in the periodic table, with the result that cadmium can replace zinc in many vital enzymes. Similarities can also occur between elements lying in adjacent positions in rows of the periodic table. For example, platinum lies next to gold. It has long been known that an inorganic compound of platinum called cisplatin can cure various forms of cancer. As a result, many drugs have been developed in which gold atoms are made to take the place of platinum, and this has produced some successful new drugs.

One final example of the medical consequences of how elements are placed in the periodic table is provided by rubidium, which lies directly below potassium in group 1 of the table. As in the previous cases mentioned, atoms of rubidium can mimic those of potassium, and so like potassium can easily be absorbed into the human body. This behaviour is exploited in monitoring techniques, since rubidium is attracted to cancers, especially those occurring in the brain.

The conventional periodic table consists of rows and columns. Trends can be observed among the elements going across and down the table. Each horizontal row represents a single period of the table. On crossing a period, one passes from metals such as potassium and calcium on the left, through transition metals such as iron, cobalt, and nickel, then through some metalloid elements like germanium, and on to some non-metals such as arsenic, selenium, and bromine, on the right side of the table. In general, there is a smooth gradation in chemical and physical properties as a period is crossed, but exceptions to this general rule abound and make the study of chemistry a fascinating and unpredictably complex field.

Metals themselves can vary from soft dull solids like sodium or potassium to hard shiny substances like chromium, platinum, or iron. Non-metals, on the other hand, tend to be solids or gases, such as carbon and oxygen respectively. In terms of their appearance, it is sometimes difficult to distinguish between solid metals and solid non-metals. To the layperson, a hard and shiny non-metal like boron may seem to be more metallic than a soft metal like sodium. The periodic trend from metals to non-metals is repeated with each period, so that when the rows are stacked, they form columns, or groups, of similar elements. Elements within a single group tend to share many important physical and chemical properties, although there are many exceptions.

In 1990, the IUPAC recommended that groups should be sequentially numbered with an Arabic numeral, in place of a Roman numeral, from left to right, as groups 1 to 18 (Figure 4), without the use of the letters A or B that can be seen in older periodic tables.

H																	He
Li	Be											B	C	N	O	F	Ne
Na	Mg											Al	Si	P	S	Cl	Ar
K	Ca	Sc	Ti	V	Cr	Mn	Fe	Co	Ni	Cu	Zn	Ga	Ge	As	Se	Br	Kr
Rb	Sr	Y	Zr	Nb	Mo	Tc	Ru	Rh	Pd	Ag	Cd	In	Sn	Sb	Te	I	Xe
Cs	Ba	Lu	Hf	Ta	W	Re	Os	Ir	Pt	Au	Hg	Tl	Pb	Bi	Po	At	Rn
Fr	Ra	Lr	Rf	Db	Sg	Bh	Hs	Mt	Ds	Rg	Cn	Nh	Fl	Mc	Lv	Ts	Og

La	Ce	Pr	Nd	Pm	Sm	Eu	Gd	Tb	Dy	Ho	Er	Tm	Yb
Ac	Th	Pa	U	Np	Pu	Am	Cm	Bk	Cf	Es	Fm	Md	No

4. Medium-long-form table.

Forms of the periodic table

There have been quite literally over 1,000 periodic tables published in print, or more recently on the Internet. How are they all related? Is there one optimal periodic table? These are questions that will be explored in this book since they can teach us many interesting things about modern science.

One aspect of this question should be addressed right away. One of the ways of classifying the periodic tables that have been published is to consider three basic formats. First of all, there are the originally produced short-form tables published by the pioneers of the periodic table like Newlands, Lothar Meyer, and Mendeleev, all of which will be examined more closely in due course (Figure 5).

These tables essentially crammed all the then known elements into eight vertical columns or groups. They reflect the fact that, broadly speaking, the elements seem to recur after an interval of eight elements if the elements are arranged in a natural sequence (another topic to be discussed). As more information was gathered on the properties of the elements, and as more elements were discovered, a new kind of arrangement called the medium-long-form table (Figure 4) began to gain prominence. Today, this form is almost completely ubiquitous. One odd feature is that the main body of the table does not contain all the elements. If you look at Figure 4, you will find a break between elements 56 and 71, and then again between elements 88 and 103. The 'missing' elements are grouped together in what looks like a separate footnote that lies below the main table.

This act of separating off the lanthanoid and actinoid elements is performed purely for convenience. If it were not carried out, the periodic table would appear much wider, thirty-two elements wide to be precise, instead of eighteen elements wide. The thirty-two-wide format does not lend itself readily to being reproduced on the inside cover of chemistry textbooks or on large wall-charts that hang in lecture rooms and laboratories. But if the elements are shown in this expanded form, as they sometimes are, one has the long-form periodic table, which may be said to be more correct than the familiar medium-long form, in the sense that the sequence of elements is unbroken.

But what are the occupants of the periodic table? Let us go back to the periodic table in general and let us choose the familiar medium-long format in order to put some flesh onto the skeleton, or framework, provided by this two-dimensional grid. How were the elements discovered? What are the elements like? How do they differ as we move down a column of the periodic table or across a horizontal period?

Typical groups of elements in the periodic table

On the extreme left of the table, group 1 contains such elements as the metals sodium, potassium, and rubidium. These are unusually soft and reactive substances, quite unlike what are normally considered as metals, such as iron, chromium, gold, or silver. The metals of group 1 are so reactive that merely placing a small piece of one of them into water gives rise to a vigorous reaction that produces hydrogen gas and leaves behind a colourless alkaline solution. The elements in group 2 include magnesium, calcium, and barium, and tend to be less reactive than those of group 1 in most respects.

Moving to the right, one encounters a central rectangular block of elements collectively known as the transition metals, which includes such examples as chromium, nickel and copper, and iron. In early periodic tables, known as short-form tables (Figure 5), these elements were placed among the groups of what are now called the main group elements.

Several valuable features of the chemistry of these elements are lost in the modern table because of the manner in which they have been separated from the main body, although the advantages of this later organization outweigh these losses. To the right of the transition metals, in the medium-long-form table, lies another block of representative elements starting with group 13 and ending with group 18, the noble gases on the extreme right of the table.

Sometimes the properties a group shares are not immediately obvious. This is the case with group 14, which consists of carbon, silicon, germanium, tin, and lead. Here, one notices a great diversity on progressing down the group. Carbon, at the head of the group, is a non-metal solid that occurs in three completely different structural forms (diamond, graphite, and fullerenes), and which forms the basis of all living systems. The next element below, silicon, is a metalloid which, interestingly, forms the basis of artificial life, or at least artificial intelligence, since it lies at the heart of all computers. The next element down, germanium, is a more recently discovered metalloid that was predicted by Mendeleev and later found to have many of the properties he

MENDELÉEFF'S TABLE I.—1871.

Series.	GROUP I. R_2O.	GROUP II. RO.	GROUP III. R_2O_3.	GROUP IV. RH_4. RO_2.	GROUP V. RH_3. R_2O_5.	GROUP VI. RH_2. RO_3.	GROUP VII. RH. R_2O_7.	GROUP VIII. RO_4.
1	H=1							
2	Li=7	Be=9.4	B=11	C=12	N=14	O=16	F=19	
3	Na=23	Mg=24	Al=27.3	Si=28	P=31	S=32	Cl=35.5	
4	K=39	Ca=40	—=44	Ti=48	V=51	Cr=52	Mn=55	Fe=56, Ce=59 Ni=59, Cu=63
5	(Cu=63)	Zn=65	—=68	—=72	As=75	Se=78	Br=80	
6	Rb=85	Sr=87	? Y=88	Zr=90	Nb=94	Mo=96	—=100	Ru=104, Rh=104 Pd=106, Ag=108
7	(Ag=108)	Cd=112	In=113	Sn=118	Sb=122	Te=125	I=127	
8	Cs=133	Ba=137	? Di=138	? Ce=140	....	....	....	....
9	....	....	....	....	....	....	....	....
10	....	....	? Er=178	? La=180	Ta=182	W=184	....	Os=195, In=197 Pt=198, Au=199
11	(Au=199)	Hg=200	Tl=204	Pb=207	Bi=208	....	....	
12	....	....	....	Th=231	....	U=240	....	....

5. Short-form periodic table, published by Mendeleev in 1871.

foresaw. On moving down to tin and lead, one arrives at two metals known since antiquity. In spite of this wide variation among them, in terms of metal–non-metal behaviour, the elements of group 14 nevertheless are similar in an important chemical sense in that they all display a maximum valency of four, meaning they all have the ability to form four bonds.

The apparent diversity of the elements in group 17 is even more pronounced. The elements fluorine and chlorine, which head the group, are both poisonous gases. The next member, bromine, is one of the only two known elements that exist as liquids at room temperature, the other one being the metal mercury. Moving further down the group, one then encounters iodine, a violet-black solid element. If a novice chemist were asked to group these elements according to their appearances, it is unlikely that he or she would consider classifying together fluorine, chlorine, bromine, and iodine. This is one instance where the subtle distinction between the observable and the abstract senses of the concept of an element can be helpful. The similarity between them lies primarily in the nature of the abstract elements and not the elements as substances that can be isolated and observed, an issue that has been discussed by Fritz Paneth (see Further reading).

On moving all the way to the right, a remarkable group of elements is encountered. These are the noble gases, all of which were first isolated just before or at the turn of the 20th century. Their main property, rather paradoxically, at least when they were first isolated, was that they lacked chemical properties. These elements, consisting of helium, neon, argon, and krypton, were not even included in early periodic tables, since they were unknown and unanticipated. When they were discovered, their existence posed a formidable challenge to the periodic system, but one that was eventually accommodated successfully by the extension of the table to include a new group, now labelled as group 18.

Another block of elements, found at the foot of the modern table, consists of the lanthanoids and actinoids that are commonly depicted as being literally disconnected. But this is just an apparent feature of this generally used display of the periodic system. Just as the transition metals are generally inserted as a block into the main body of the table, it is quite possible to do the same with the lanthanoids and actinoids. Indeed, many such long-form displays have been published. While the long-form tables (Figure 6) give these elements a more natural place among the rest of the elements, they are rather cumbersome and do not readily lend themselves to conveniently shaped wall-charts of the periodic system. Although there are a number of different forms of the periodic table, what underlies the entire edifice, no matter the form of its representation, is the periodic law.

The periodic law

The periodic law states that after certain regular but varying intervals, the chemical elements show an approximate repetition in their properties. For example, fluorine, chlorine, and bromine, which all fall into group 17, share the property of forming white crystalline salts of formula NaX with the metal sodium (where X is any halogen atom). This periodic repetition of properties is the essential fact that underlies all aspects of the periodic system.

This talk of the periodic law raises some interesting philosophical issues. First of all, periodicity among the elements is neither constant nor exact. In the generally used medium-long form of the periodic table, the first row has two elements, the second and third each contain eight, the fourth and fifth contain eighteen, and so on. This implies a varying periodicity consisting of 2, 8, 8, 18, etc., quite unlike the kind of periodicity one finds in the days of the week or notes in a musical scale. In these other cases, the period length is constant, such as seven for the days of the week as well as the number of notes in a Western musical scale.

H 1																															He 2
Li 3	Be 4																									B 5	C 6	N 7	O 8	F 9	Ne 10
Na 11	Mg 12																									Al 13	Si 14	P 15	S 16	Cl 17	Ar 18
K 19	Ca 20															Sc 21	Ti 22	V 23	Cr 24	Mn 25	Fe 26	Co 27	Ni 28	Cu 29	Zn 30	Ga 31	Ge 32	As 33	Se 34	Br 35	Kr 36
Rb 37	Sr 38															Y 39	Zr 40	Nb 41	Mo 42	Tc 43	Ru 44	Rh 45	Pd 46	Ag 47	Cd 48	In 49	Sn 50	Sb 51	Te 52	I 53	Xe 54
Cs 55	Ba 56	La 57	Ce 58	Pr 59	Nd 60	Pm 61	Sm 62	Eu 63	Gd 64	Tb 65	Dy 66	Ho 67	Er 68	Tm 69	Yb 70	Lu 71	Hf 72	Ta 73	W 74	Re 75	Os 76	Ir 77	Pt 78	Au 79	Hg 80	Tl 81	Pb 82	Bi 83	Po 84	At 85	Rn 86
Fr 87	Ra 88	Ac 89	Th 90	Pa 91	U 92	Np 93	Pu 94	Am 95	Cm 96	Bk 97	Cf 98	Es 99	Fm 100	Md 101	No 102	Lr 103	Rf 104	Db 105	Sg 106	Bh 107	Hs 108	Mt 109	Ds 110	Rg 111	Cn 112	Nh 113	Fl 114	Mc 115	Lv 116	Ts 117	Og 118

6. Long-form periodic table.

Among the elements, however, not only does the period length vary, but the periodicity is not exact. The elements within any column of the periodic table are not exact recurrences of each other. In this respect, their periodicity is not unlike the musical scale, in which one returns to a note denoted by the same letter, which sounds like the original note but is definitely not identical to it in that it is an octave higher.

The varying length of the periods of elements and the approximate nature of the repetition has caused some chemists to abandon the term 'law' in connection with chemical periodicity. Chemical periodicity may not seem as law-like as most laws of physics. However, it can be argued that chemical periodicity offers an example of a typically chemical law: approximate and complex, but still fundamentally displaying law-like behaviour.

Perhaps this is a good place to discuss some other points of terminology. How is a periodic table different from a periodic system? The term 'periodic system' is the more general of the two. The periodic system is the more abstract notion that holds that there is a fundamental relationship among the elements. Once it becomes a matter of displaying the periodic system, one can choose a three-dimensional arrangement, a circular shape, or any number of different two-dimensional tables. Of course, the term 'table' strictly implies a two-dimensional representation. So although the term 'periodic table' is by far the best known of the three terms—law, system, and table—it is actually the most restricted.

Reacting elements and ordering the elements

Much of what is known about the elements has been learned from the way they react with other elements and from their bonding properties. The metals on the left-hand side of the conventional periodic table are the complementary opposites of the non-metals, which tend to lie towards the right-hand side. This is so because, in modern terms, metals form positive ions by the loss of electrons, while non-metals gain electrons to form negative ions. Such oppositely charged ions combine together to form electrically neutral salts like sodium chloride or calcium bromide. There are further complementary aspects of metals and non-metals. Metal oxides or hydroxides dissolve in water to form bases, while non-metal oxides or hydroxides dissolve in water to form acids. An acid and a base react together in a 'neutralization' reaction to form a salt and water. Bases and acids, just like the metals and non-metals from which they are formed, are also opposite but complementary.

Acids and bases have a connection with the origins of the periodic system since they featured prominently in the concept of equivalent weights, which was first used to order the elements. The equivalent weight of any particular metal, for example, was originally obtained from the mass of metal that reacts with a certain mass of a chosen standard acid. The term 'equivalent weight' was subsequently generalized to denote the amount of an element that reacts with a standard amount of oxygen. Historically, the ordering of the elements across periods was determined by equivalent weight, then later by atomic weight, and eventually by atomic number (explained below).

Chemists first began to make quantitative comparisons among the amounts of acids and bases that reacted together. This procedure was then extended to reactions between acids and metals. This allowed chemists to order the metals on a numerical scale according to their equivalent weight, which, as mentioned, is just the mass of the metal that combines with a fixed mass of acid.

Atomic weights, as distinct from equivalent weights, were first obtained in the early 1800s by John Dalton, who indirectly inferred them from measurements on the masses of elements combining together. But there were complications in this apparently simple method that forced Dalton to make assumptions about the chemical formulas of the compounds in question. The key to this question is the valence, or combining power, of an element. For example, a univalent atom combines with hydrogen atoms in a ratio of 1:1. Divalent atoms, like oxygen, combine in a ratio of 2:1, and so on.

Equivalent weight, as already mentioned, is sometimes regarded as a purely empirical concept since it does not seem to depend upon whether one believes in the existence of atoms. Following the introduction of atomic weights, many chemists who felt uneasy about the notion of atoms attempted to revert to the older concept of equivalent weights. They believed that equivalent weights would be purely empirical and therefore more reliable. But such hopes were an illusion, since equivalent weights also rested on the assumption of particular formulas for compounds, and formulas are theoretical notions.

For many years, there was a great deal of confusion created by the alternative use of equivalent weight and atomic weight. Dalton himself assumed that water consisted of one atom of hydrogen combined with one atom of oxygen, which would make its atomic weight and equivalent weight the same, but his guess at the valence of oxygen turned out to be incorrect. Many authors used the terms 'equivalent weight' and 'atomic weight' interchangeably, thus further adding to the confusion. The true relationship between equivalent weight, atomic weight, and valency was only clearly established in 1860 at the first major scientific conference, which was held in Karlsruhe, Germany. This clarification and the general adoption of consistent atomic weights cleared the path for the independent discovery of the periodic system by as many as six individuals in various countries, who each proposed forms of the periodic table that were successful to varying degrees. Each of them placed the elements generally in order of increasing atomic weight.

The third, and most modern, of the ordering concepts mentioned earlier is atomic number. Once atomic number was understood, it displaced atomic weight as the ordering principle for the elements. No longer dependent on combining weights in any way, atomic number can be given a simple microscopic interpretation in terms of the structure of the atoms of any element. The atomic number of an element is given by the number of protons, or units of positive charge, in the nucleus of any of its atoms. Thus each element on the periodic table has one more proton than the element preceding it. Since the number of neutrons in the nucleus also tends to increase as one moves through the periodic table, this makes atomic number and atomic weights roughly correspondent, but it is the atomic number that identifies any particular element. In other words, atoms of any particular element always have the same number of protons but they can differ in the number of neutrons they contain, a feature that produces the phenomenon of isotopy. These variants are called *isotopes*.

Different representations of the periodic system

The modern periodic system succeeds remarkably well in ordering the elements by atomic number in such a way that they fall into natural groups, but this system can be represented in more than one way. Thus there are many forms of the periodic table, some designed for different uses. Whereas a chemist might favour a form that highlights the reactivity of the elements, an electrical engineer might wish to focus on similarities and patterns in electrical conductivities.

The way in which the periodic system is displayed is a fascinating issue, and one that especially appeals to the popular imagination. Since the time of the early periodic tables of Newlands, Lothar Meyer, and Mendeleev, there have been many attempts to obtain the 'ultimate' periodic table. Indeed, it has been estimated that within one hundred years of the introduction of Mendeleev's famous table of 1869, approximately 700 different versions of the periodic table had been published. They included all kinds of alternatives, such as three-dimensional tables, helices, concentric circles, spirals, zigzags, step tables, mirror-image tables, and so on. Even today, articles are regularly published purporting to show new and improved versions of the periodic system.

What is fundamental to all these attempts is the periodic *law* itself, which exists in only one form. None of the multitude of displays changes this aspect of the periodic system. Many chemists stress that it does not matter how this law is physically represented, provided that certain basic requirements are met. Nevertheless, from a philosophical point of view, it is still relevant to consider the most fundamental representation of the elements, or the ultimate form of the periodic system, especially as this relates to the question of whether the periodic law should be regarded in a realistic manner or as a matter of convention. The usual response—that representation is only a matter of convention—would seem to clash with the realist notion that there may be a fact of the matter concerning the points at which the repetitions in properties occur in any periodic table.

Changes in the periodic table

In 1945, the American chemist Glen Seaborg suggested that the elements beginning with actinium, number 89, should be considered as an analogous series to the rare earth series, whereas it had previously been supposed that the new series would begin after element number 92, or uranium (Figure 7). Seaborg's new periodic table revealed an analogy between europium (63) and gadolinium (64) and the as yet undiscovered elements 95 and 96 respectively. On the basis of these analogies, Seaborg succeeded in synthesizing and identifying the two new elements, which were subsequently named americium and curium. More than twenty further transuranium elements have subsequently been synthesized.

																H	He
Li	Be											B	C	N	O	F	Ne
Na	Mg											Al	Si	P	S	Cl	Ar
K	Ca	Sc	Ti	V	Cr	Mn	Fe	Co	Ni	Cu	Zn	Ga	Ge	As	Se	Br	Kr
Rb	Sr	Y	Zr	Nb	Mo	Tc	Ru	Rh	Pd	Ag	Cd	In	Sn	Sb	Te	I	Xe
Cs	Ba	RE	Hf	Ta	W	Re	Os	Ir	Pt	Au	Hg	Tl	Pb	Bi	Po	At	Rn
Fr	Ra	AC	Th	Pa	U												

rare earths	La	Ce	Pr	Nd	Pm	Sm	Eu	Gd	Tb	Dy	Ho	Er	Tm	Yb	Lu

																H	He
Li	Be											B	C	N	O	F	Ne
Na	Mg											Al	Si	P	S	Cl	Ar
K	Ca	Sc	Ti	V	Cr	Mn	Fe	Co	Ni	Cu	Zn	Ga	Ge	As	Se	Br	Kr
Rb	Sr	Y	Zr	Nb	Mo	Tc	Ru	Rh	Pd	Ag	Cd	In	Sn	Sb	Te	I	Xe
Cs	Ba	LA	Hf	Ta	W	Re	Os	Ir	Pt	Au	Hg	Tl	Pb	Bi	Po	At	Rn
Fr	Ra	AC															

La	La	Ce	Pr	Nd	Pm	Sm	Eu	Gd	Tb	Dy	Ho	Er	Tm	Yb	Lu
Ac	Ac	Th	Pa	U	Np	Pu									

7. Periodic tables before and after Seaborg's modification.

The standard form of the periodic table has also undergone some minor changes regarding the elements that mark the beginning of the third and fourth periods of the transition elements. Whereas older periodic tables show these elements to be lanthanum (57) and actinium (89), more recent experimental evidence and analysis have put lutetium (71) and lawrencium (103) in their former places (see Chapter 10). It is also interesting to note that some even older periodic tables based on macroscopic properties had anticipated these changes.

These are examples of ambiguities in what may be termed secondary classification, which is not as unequivocal as primary classification, or the sequential ordering of the elements. In classical chemical terms, secondary classification corresponds to the chemical similarities between the various elements in a group. In modern terms, secondary classification is explained using the concept of electronic configurations. Regardless of whether one takes a classical chemical approach or a more physical approach based on electronic configurations, secondary classification of this type is more tenuous than primary classification and cannot be established as categorically. The way in which secondary classification, as defined here, is established is a modern example of the tension between using chemical properties or physical properties for classification. The precise placement of an element within groups of the periodic table can vary depending on whether one puts more emphasis on electronic configuration (a physical property) or its chemical properties. In fact, many recent debates on the placement of helium in the periodic system revolve around the relative importance that should be assigned to these two approaches (see Chapter 10).

In recent years, the number of elements has increased well beyond a hundred as the result of the synthesis of artificial elements. At the time of writing, all the elements up to and including element 118 have been synthesized and characterized. Such elements are typically very unstable and only a few atoms are produced at any time. However, ingenious chemical techniques have been devised that permit the chemical properties of these so-called 'superheavy' elements to be examined and allow one to check whether extrapolations of chemical properties are maintained for such highly massive atoms.

On a more philosophical note, the production of these elements allows us to examine whether the periodic law is an exceptionless law, of the same kind as Newton's law of gravitation, or whether deviations from the expected recurrences in chemical properties might take place once a sufficiently high atomic number is reached. No surprises have been found so far, but the question of whether some of these superheavy elements have the expected chemical properties is far from being fully resolved. One important complication that arises in this region of the periodic table is the increasing significance of relativistic effects (see next section). These effects cause the adoption of unexpected electronic configurations in some atoms and may result in equally unexpected chemical properties.

Understanding the periodic system

Developments in physics have had a profound influence on the manner in which the periodic system is now understood. The two most important theories in modern physics are Einstein's theory of relativity and quantum mechanics.

The first of these has had a limited impact on our understanding of the periodic system but is becoming increasingly important in accurate calculations carried out on atoms and molecules. The need to take account of relativity arises whenever objects move at speeds close to that of light. Inner electrons, especially those in the heavier atoms in the periodic system, can readily attain such relativistic velocities. In order to carry out accurate calculations, especially on heavy atoms, it is essential to introduce relativistic corrections. In addition, many seemingly mundane properties of elements, like the characteristic colour of gold or the liquidity of mercury, can best be explained as relativistic effects due to fast-moving inner-shell electrons.

But it is the second theory of modern physics that has exerted by far the more important role in attempts to understand the periodic system theoretically. Quantum theory was actually born in the year 1900. It was first applied to atoms by Niels Bohr, who pursued the notion that the similarities between the elements in any group of the periodic table could be explained by their having equal numbers of outer-shell electrons. The very notion of a particular number of electrons in an electron shell is an essentially quantum-like concept. Electrons are associated with certain quanta, or packets, of energy and to therefore lie in one or another shell around the nucleus of the atom (see Chapter 7).

Soon after Bohr had introduced the quantum to the atom, many others developed his theory until the old quantum theory gave rise to quantum mechanics (see Chapter 8). Under the new description, electrons are regarded as much as waves as they are as particles. Even stranger is the notion that electrons no longer follow definite trajectories, or orbits, around the nucleus. Instead, the description changes to talk of smeared-out electron clouds, which occupy so-called orbitals. The most recent explanation of the periodic system is given in terms of how many such orbitals are populated by electrons. The explanation depends on the electron arrangement, or 'configuration', of an atom, which is spelled out in terms of the occupation of its orbitals.

The interesting question raised here is the relationship between chemistry and modern atomic physics, and in particular quantum mechanics. The popular view reinforced in most textbooks is that chemistry is nothing but physics 'deep down' and that all chemical phenomena, and especially the periodic system, can be developed on the basis of quantum mechanics. There are some problems with this view, however, which will be considered. For example, it will be suggested that the quantum mechanical explanation for the periodic system is still far from perfect. This is important because chemistry books, especially textbooks aimed at teaching, tend to give the impression that our current explanation of the periodic system is essentially complete. This is just not the case, as will be discussed.

The periodic system ranks as one of the most fruitful and unifying ideas in the whole of modern science, comparable perhaps with Darwin's theory of evolution by natural selection. After evolving for nearly 150 years through the work of numerous individuals, the periodic table remains at the heart of the study of chemistry. This is mainly because it is of immense practical benefit for making predictions about all manner of chemical and physical properties of the elements and possibilities for bond formation. Instead of having to learn the properties of the one hundred or more elements, the modern chemist, or the student of chemistry, can make effective predictions from knowing the properties of typical members of each of the eight main groups and those of the transition metals and lanthanoids and actinoids elements.

Having laid some thematic foundations and defined some key terms, we will now begin the story of the development of the modern periodic system, starting with its birth in the 18th and 19th centuries.

Chapter 3
Atomic weight, triads, and Prout

The first classifications of elements into related families were carried out on the basis of chemical similarities between elements, that is to say, on the basis of qualitative rather than quantitative aspects of the elements. For example, it was clear that the metals lithium, sodium, and potassium shared many similarities including their being soft, their floating on water, and the fact that unlike most metals, they could react visibly with water.

But the modern periodic table is based as much on quantitative properties of the elements as it is on their qualitative ones. Chemistry as a whole began to be a quantitative field, meaning *how much* reacts rather than *how* it reacts, as early as the 16th and 17th centuries. One of those responsible for this approach was Antoine Lavoisier, a French nobleman who was later guillotined in the aftermath of the French Revolution. Lavoisier was among the first to make accurate measurements of the weights of chemical reactants and their products. By doing this, Lavoisier was able to refute a long-standing assumption that a substance called 'phlogiston' was evolved when substances were burned.

On the contrary, Lavoisier discovered that burning any substance, such as an element, resulted in an increase rather than a decrease in weight. He also discovered that in any chemical operation, an equal quantity of matter exists before and after the operation.

The discovery of this 'law of conservation of matter' was followed by other laws of chemical combination, all of which began to demand a deeper explanation and one that would eventually lead to the formulation of the periodic table.

Lavoisier also turned away from the Greek notion of an abstract element as a bearer of properties. Instead, he concentrated on elements as the final stage in the decomposition of any compound. Although the notion of abstract elements would later return, in a modified form, it was necessary to make a clean break with the ancient Greek tradition, especially as many mysterious and non-scientific notions had continued to flourish in the Middle Ages and among the alchemists.

Returning to quantitative aspects of the elements, Benjamin Richter, working in Germany in 1792, published a list of what became known as equivalent weights (Figure 8). This was a list of

Base		***Acid***	
Alumina	525	Hydrofluoric	427
Magnesia	615	Carbonic	577
Ammonia	672	Sebacic	706
Lime	793	Muriatic	712
Soda	859	Oxalic	755
Strontia	1,329	Phoshproic	979
Potash	1,605	Sulphuric	1,000
Baryta	2,222	Succinic	1,209
		Nitric	1,405
		Acetic	1,480
		Citric	1,583
		Tartaric	1,694

8. Richter's table of equivalent weights, as modified by Fischer in 1802.

the weights of various metals that reacted with a fixed amount of a particular acid, nitric acid for example. Now for the first time, the properties of various elements could be compared with each other in a simple quantitative way.

Dalton

In 1801, a young schoolteacher in Manchester, England, published the beginnings of a modern atomic theory. Following in the new tradition of Lavoisier and Richter, John Dalton took the ancient Greek notion of atoms, or the smallest particles of any substance, and made it quantitative. Not only did he assume that each element consisted of a particular type of atom, but he also began to estimate their relative weights.

For example, he drew on Lavoisier's experiments on the combination of hydrogen and oxygen to form water. Lavoisier had shown that water was composed of 85 per cent oxygen and 15 per cent hydrogen by mass. Dalton proposed that water consisted of one atom of hydrogen combined with one of oxygen to give a formula of HO, and that the atomic weight of oxygen was therefore 85/15 = 5.66, assuming the weight of a hydrogen atom to be one unit. The current value for the oxygen atom is actually 16, and the difference lies in two issues that Dalton was unaware of. First of all, he was incorrect to assume that water is HO since, as everybody now knows, the formula is rather H_2O. Second, Lavoisier's data were not very accurate.

Dalton's notion of atomic weights gave a very reasonable explanation for the law of constant proportion, namely the fact that when two elements combine together, they do so in a constant ratio of their weights. This law could now be regarded as a scaled-up version of the combination of two or more atoms having particular atomic weights. The fact that macroscopic samples consist of a fixed ratio by weight of two elements reflects the fact that two particular atoms are combining many times over and, since they have particular masses, the product will also reflect that mass ratio.

Other chemists, as well as Dalton himself, discovered yet another law of chemical combination, the 'law of multiple proportions'. When one element, A, combines with another one, B, to form more than one compound, there is a simple ratio between the combining masses of B in the two compounds. For example, carbon and oxygen combine together to form carbon monoxide and carbon dioxide. The weight of combined oxygen in the dioxide is twice as much as the weight of combined oxygen in the monoxide. Again, this law found a good explanation in Dalton's atomic theory because it implied that one atom of carbon had combined with one atom of oxygen in carbon monoxide, CO, and two atoms of oxygen had combined with one of carbon in carbon dioxide, CO_2.

Von Humboldt and Gay-Lussac

Now consider one further law of chemical combination, but one which was not at first explained by Dalton's theory. In 1809, Alexander von Humboldt and Joseph Louis Gay-Lussac discovered that when water vapour was formed by reacting two gases, hydrogen and oxygen, the volume of hydrogen consumed was almost twice that of oxygen. Moreover, the volume of water vapour formed was about the same as that of the combining hydrogen.

> 2 volumes of hydrogen + 1 volume of oxygen → 2 volumes of water vapour

This kind of behaviour was also found to apply to other gases combining together, so that von Humboldt and Gay-Lussac could conclude:

> The volumes of gases entering into chemical reaction and the gaseous products are in a ratio of small integers.

This new law of chemistry presented a major challenge to Dalton's new atomic theory. According to Dalton, any atom was absolutely indivisible, but this law could not be interpreted on the assumption of indivisible atoms of the gases concerned. It is only if the oxygen atoms are divisible that the above reaction between hydrogen and oxygen can possibly take place between two atoms of hydrogen and one of oxygen.

The solution to this puzzle came when an Italian physicist, Amedeo Avogadro, realized that it was rather two diatomic molecules of hydrogen that were combining with one diatomic molecule of oxygen. Nobody had previously thought that these gases consisted of two atoms of the element combined together to form a diatomic molecule. Since these molecules consisted of two atoms, it was the molecules that were divisible not the atoms themselves. Dalton's theory and the indivisibility of atoms could still be upheld and, by assuming the existence of diatomic gas molecules, consisting of two atoms of the same element, the new law of von Humboldt and Gay-Lussac could also be explained.

The reaction between hydrogen and oxygen was such that two hydrogen diatomic molecules broke up to form four atoms while one oxygen diatomic molecule broke up to form two oxygen atoms. Two molecules of water vapour, or H_2O, would then form to account for all the six atoms in question. This all looks very simple in hindsight, but given that diatomic molecules represented a radical idea and that the formula for the water molecule was not known, it is not surprising that an equation as simple as,

$$2H_2 + O_2 \rightarrow 2H_2O$$

took about fifty years before it was fully understood.

But by an odd historical twist, Dalton himself refused to accept the idea of diatomic molecules because he firmly believed that any two atoms of the same element should repel each other and that as a result they could never form a diatomic molecule. The idea of a chemical bond between any two like atoms was new and took some getting used to, especially to somebody like Dalton who had a rather elaborate view of how atoms should behave. Meanwhile, somebody like Avogadro could forge ahead and postulate diatomic molecules while being unencumbered by the idea that two like atoms repel each other, which in fact they do not, as we realize today.

Avogadro's idea of the formation of diatomic molecules was also arrived at independently by André Ampère, after whom the ampere, or unit of current, is now named. But this crucial discovery lay dormant for something like fifty years until it was finally resuscitated by another Italian, Stanislao Cannizzaro, who lived in Sicily.

Prout's hypothesis

A few years after Dalton and others had begun publishing lists of atomic weights, the Scottish physician William Prout noticed something rather interesting. Many of the atomic weights that had been determined for the elements seemed to be whole-number multiples of the weight of the hydrogen atom. His conclusion was rather an obvious one. Perhaps all the atoms were simply composed of atoms of hydrogen. If this was true, it would also suggest the unity of all matter at the fundamental level, a notion that has been toyed with since the dawn of Greek philosophy and has resurfaced many times in different forms.

But not all the published atomic weights were exact multiples of the weight of hydrogen. Prout was not put off by this feature but suggested that the cause lay in the fact that the weights of these aberrant atoms had not yet been accurately determined. Prout's hypothesis, as it came to be known, was therefore rather productive because it caused others to measure atomic weights more and more accurately in order to prove him either right or wrong. The ensuing more accurate atomic weights would eventually play a pivotal role in the discovery and evolution of the periodic table.

But the initial consensus regarding Prout's hypothesis was that it was incorrect. The more accurately measured atomic weights suggested that, in general, atoms were not multiples of hydrogen atoms. Nevertheless, Prout's hypothesis was destined to make a comeback a good deal later in the story, although in a somewhat modified form.

Döbereiner's triads

The German chemist Wolfgang Döbereiner discovered another general principle that also contributed to the need to measure atomic weights more accurately and so paved the way for the periodic table. Starting in 1817, Döbereiner identified the existence of various groups of elements in which one of the elements had both chemical properties and an atomic weight which was roughly the average of two other elements. These groups of three elements became known as triads. For example, lithium, sodium, and potassium are all softish grey metals with low densities. Lithium shows little reaction with water, whereas potassium is highly reactive. But sodium shows intermediate reactivity between that of these two other members of the triad.

In addition, the atomic weight of sodium (23) is intermediate between those of lithium (7) and potassium (39). This discovery was highly significant because it gave the first hint of a numerical regularity lying at the heart of the relationship between the nature and properties of the elements. It suggested an underlying mathematical order as to how the elements were related to each other chemically.

Another key triad recognized by Döbereiner consisted of three halogen elements, chlorine, bromine, and iodine. But Döbereiner did not attempt to connect these different triads together in any way. Had he done so, he might have discovered the periodic table some fifty years before Mendeleev and others did.

In identifying several triads, Döbereiner demanded that there should be a chemical similarity among the three elements in question in addition to the mathematical relationship mentioned. Other authors who followed him were not as fussy on the first point, and some of them believed that they had discovered many other triads. For example, in 1857 a twenty-year-old German chemist, Ernst Lenssen, working in Wiesbaden, published an article in which he arranged virtually all of the fifty-eight known elements into a total of twenty triads. Ten of his triads consisted of non-metals and acid-forming metals and the remaining ten of just metals.

Using the twenty triads in Figure 9, Lenssen also claimed to identify a total of super-triads, in which the mean equivalent weight of each middle triad lies approximately midway between the mean weights of the other triads in a group of three triads. These were triads of triads, as it were. But Lenssen's system was somewhat forced. For example, in place of a genuine triad, he counted just one element, hydrogen, as constituting a triad because he found it convenient to do so. Furthermore, many of the triads that he claimed to exist looked appealing in numerical terms, but had no chemical significance. Lenssen and some other chemists became seduced by apparent numerical regularities while leaving chemistry behind.

Another system of classification of the elements was due to Leopold Gmelin working in Germany in 1843. This author discovered some new triads and did begin to connect them together to form an overall classification system with a rather distinctive shape (Figure 10). His system contained as many as fifty-five elements, and in anticipation of later systems, it appears that Gmelin ordered most of the elements in order of increasing atomic weights, although he never explicitly expressed this notion.

	Calculated atomic weight			Determined atomic weights		
1	(K + Li)/2	= Na	= 23.03	39.11	23.00	6.95
2	(Ba + Ca)/2	= Sr	= 44.29	68.59	47.63	20
3	(Mg + Cd)/2	= Zn	= 33.8	12	32.5	55.7
4	(Mn + Co)/2	= Fe	= 28.5	27.5	28	29.5
5	(La + Di)/2	= Ce	= 48.3	47.3	47	49.6
6	Yt Er Tb			32	?	?
7	Th norium Al			59.5	?	13.7
8	(Be + Ur)/2	= Zr	= 33.5	7	33.6	60
9	(Cr + Cu)/2	= Ni	= 29.3	26.8	29.6	31.7
10	(Ag + Hg)/2	= Pb	= 104	108	103.6	100
11	(O + C)/2	= N	= 7	8	7	6
12	(Si + Fl)/2	= Bo	= 12.2	15	11	9.5
13	(Cl + J)/2	= Br	= 40.6	17.7	40	63.5
14	(S + Te)/2	= Se	= 40.1	16	39.7	64.2
15	(P + Sb)/2	= As	= 38	16	37.5	60
16	(Ta + Ti)/2	= Sn	= 58.7	92.3	59	25
17	(W + Mo)/2	= V	= 69	92	68.5	46
18	(Pa + Rh)2	= Ru	= 52.5	53.2	52.1	51.2
19	(Os + Ir)/2	= Pt	= 98.9	99.4	99	98.5
20	(Bi + Au)/2	= Hg	= 101.2	104	100	98.4

9. The twenty triads of Lenssen.

O N H
F Cl Br I Li Na K
S Se Te Mg Ca Sr Ba
P As Sb Be Ce La
C B Bi Zr Th Al
Ti Ta W Sn Cd Zn
Mo V Cr U Mn Ni Fe
Bi Pb Ag Hg Cu
Os Ir Rh Pt Pd Au

10. Gmelin's table of elements.

However, Gmelin's system cannot be regarded as a periodic system since it does not display the repetition in the properties of the elements. In other words, the property of chemical periodicity from which the periodic table derives its name was not yet featured. Gmelin went on to use his system of elements in order to organize a textbook of chemistry consisting of 500 or so pages.

This is probably the first time that a table of elements was used as a basis for an entire book on chemistry, something that is completely standard these days, although we should recall that this was not a *periodic* table.

Kremers

A modern periodic table is much more than a collection of groups of elements showing similar chemical properties. In addition to what may be called 'vertical relationships', which embody triads of elements, a modern periodic table connects together groups of elements into an orderly sequence.

A periodic table consists of a horizontal dimension, containing dissimilar elements, as well as a vertical dimension with similar elements. The first person to consider a horizontal relationship was Peter Kremers from Cologne in Germany. He noted the regularity illustrated in Figure 11 among a short series of elements that included oxygen, sulphur, titanium, phosphorus, and selenium.

Kremers also discovered some new triads such as

$$Mg = \frac{O+S}{2}, \quad Ca = \frac{S+Ti}{2}, \quad Fe = \frac{Ti+P}{2}$$

From a modern standpoint, these triads may not seem to be chemically significant. But this is because the modern medium-long form of the periodic system fails to display secondary kinships among some elements. Sulphur and titanium both show a valency

	O	S	Ti	P	Se
Atomic weight	8	16	24.12	32	39.62
Difference	8	8	~8	~8	

11. Kremers' atomic weight differences for the oxygen series.

of four, for example, though they do not appear in the same group in the medium-long form of the periodic system. But it is not so far-fetched to consider them as being chemically analogous. Given the fact that both titanium and phosphorus commonly display valences of three, this grouping is also not as incorrect as a modern reader may think. But broadly speaking, as in the case of Lenssen, this is a desperate attempt to create new triads at all costs. The aim seems to have become one of finding triad relationships among the weights of the elements irrespective of whether or not these had any chemical significance. Mendeleev later described such activity among his colleagues as an obsession with triads, which he believed had delayed the discovery of the mature periodic system.

But to return to Kremers, his most incisive contribution lay in the suggestion of a bi-directional scheme of what he termed 'conjugated triads'. Here, certain elements would serve as members of two distinct triads lying perpendicularly to each other,

Li 6.5	Na 23	K 39.2
Mg 12	Zn 32.6	Cd 56
Ca 20	Sr 43.8	Ba 68.5

Thus, in a more profound way than any of his predecessors, Kremers was comparing chemically *dissimilar* elements, a practice which would only reach full maturity with the tables of Lothar Meyer and Mendeleev.

Chapter 4

Steps towards the periodic table

The 1860s was an important decade for the formulation of the periodic table. It began with a congress convened in Karlsruhe, Germany, aimed at resolving a number of technical issues having to do with chemists' understanding of the concept of atoms and molecules.

As mentioned in Chapter 2, the law of combining gas volumes discovered by Gay-Lussac could only be explained by assuming the existence of divisible diatomic molecules of two combined atoms, such as H_2, O_2, etc. This proposal had still not been generally accepted because of criticisms from Dalton, among others. At the Karlsruhe conference, the idea finally gathered widespread acceptance due to the advocacy of Cannizzaro, a fellow countryman of Avogadro who had first suggested the idea fifty years previously.

Another problem had been that many different authors gave different values of the atomic weights of the elements. Cannizzaro also succeeded in producing a rationalized set of values that he printed in a small pamphlet that was distributed to delegates departing from the conference. Armed with these reforms, it was just a matter of time before as many as six individual scientists developed rudimentary periodic systems which included most of the then known sixty or so elements.

De Chancourtois

The first person to actually discover chemical periodicity was a French geologist, Alexandre-Émile Béguyer de Chancourtois, who arranged the elements in order of increasing atomic weights on a spiral that he inscribed around a metal cylinder. After doing so, he noticed that chemically similar elements fell on vertical lines, which intersected the spiral as it circled the cylinder (Figure 12). Here was the essential discovery that the elements, when arranged in their natural order, seemed to recur approximately after certain regular intervals. Just like the days of the week, the months of the year, or the notes in a musical scale, periodicity or repetition seems to be an essential property of the elements. The underlying cause of chemical repetition would remain mysterious for many years to come.

De Chancourtois expressed his support for Prout's hypothesis and went as far as rounding off the values of atomic weights to whole numbers in his periodic system. Sodium, with a weight of 23, appeared in his system at one whole turn away from lithium, with an atomic weight of 7. In the next column, he placed the elements magnesium, calcium, iron, strontium, uranium, and barium. In the modern periodic system, four of these elements—magnesium, calcium, strontium, and barium—are indeed in the same group. The inclusion of iron and uranium seems at first to be an outright error, but as we will see, many early short-form periodic tables featured some transition elements among what are now termed the main group elements.

The Frenchman was also somewhat unlucky in that his first and most important publication failed to include a diagram of his system, a rather fatal omission given the fact that graphical representation is *the* crucial aspect of any periodic system. In order to remedy this problem, De Chancourtois later republished his paper privately, but as a result it did not gain widespread distribution and remained obscured from the chemical community

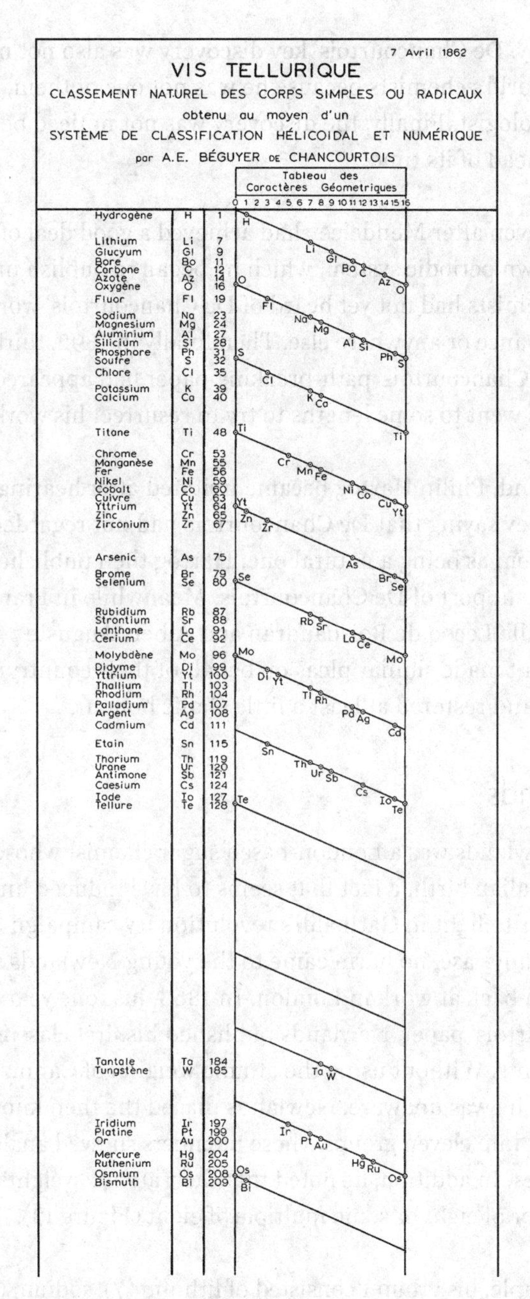

12. De Chancourtois' telluric screw.

of the day. De Chancourtois' key discovery was also not noticed by the world's chemists because he was not one of them, but was a geologist. Finally, the discovery was not noticed because it was ahead of its time.

In fact, even after Mendeleev had achieved a good deal of fame for his own periodic system, which he began to publish in 1869, most chemists had not yet heard of De Chancourtois' work, in his native France or anywhere else. Then finally in 1892, thirty years after De Chancourtois' path-breaking paper had appeared, three chemists went to some lengths to try to resurrect his work.

In England, Philip Hartog became irritated after hearing Mendeleev saying that De Chancourtois had not regarded his own system as being a natural one. Hartog then published a paper in support of De Chancourtois. Meanwhile in France, Paul-Émile Lecoq de Boisbaudran and Albert Auguste Lapparent made similar pleas on behalf of their countryman's priority and restored at least a little Gallic honour.

Newlands

John Newlands was a London-based sugar chemist whose mother was of Italian birth, a fact that seems to have induced him to volunteer to fight in Garibaldi's revolutionary campaign to unify Italy. In any case, no harm came to the young Newlands since he was soon back at work in London. In 1863, just one year after De Chancourtois' paper, Newlands published his first classification of elements. Without using the atomic weights of Cannizzaro, of which he was unaware, Newlands placed the then known elements into eleven groups whose members showed analogous properties. In addition, he noted that their atomic weights differed by a factor of eight or some multiple of eight (Figure 13).

For example, his group I consisted of lithium (7), sodium (23), potassium (39), rubidium (85), caesium (123), and thallium (204).

Group I. Metals of the alkalies :—Lithium, 7 ; sodium, 23 ; potassium, 39 ; rubidium, 85 ; cæsium, 123 ; thallium, 204.

The relation among the equivalents of this group (see CHEMICAL NEWS, January 10, 1863) may, perhaps, be most simply stated as follows :—

1 of lithium + 1 of potassium = 2 of sodium.
1 ,, + 2 ,, = 1 of rubidium.
1 ,, + 3 ,, = 1 of cæsium.
1 ,, + 4 ,, = 163, the equivalent of a metal not yet discovered.
1 ,, + 5 ,, = 1 of thallium.

Group II. Metals of the alkaline earths :—Magnesium, 12 ; calcium, 20 ; strontium, 43·8 ; barium, 68·5.

In this group, strontium is the mean of calcium and barium.

Group III. Metals of the earths :—Beryllium, 6·9 ; aluminium, 13·7 ; zirconium, 33·6 ; cerium, 47 ; lanthanium, 47 ; didymium, 48 ; thorium, 59·6.

Aluminium equals two of beryllium, or one-third of the sum of beryllium and zirconium. (Aluminium also is one-half of manganese, which, with iron and chromium, forms sesquioxides, isomorphous, with alumina.)

1 of zirconium + 1 of aluminium = 1 of cerium.
1 ,, + 2 ,, = 1 of thorium.

Lanthanium and didymium are identical with cerium, or nearly so.

Group IV. Metals whose protoxides are isomorphous with magnesia :—Magnesium, 12 ; chromium, 26·7 ; manganese, 27·6 ; iron, 28 ; cobalt, 29·5 ; nickel, 29·5 ; copper, 31·7 ; zinc, 32·6 ; cadmium, 56.

Between magnesium and cadmium, the extremities of this group, zinc is the mean. Cobalt and nickel are identical. Between cobalt and zinc, copper is the mean. Iron is one-half of cadmium. Between iron and chromium, manganese is the mean.

Group V.—Fluorine, 19 ; chlorine, 35·5 ; bromine, 80 ; iodine, 127.

In this group bromine is the mean between chlorine and iodine.

Group VI.—Oxygen, 8 ; sulphur, 16 ; selenium, 39·5 ; tellurium, 64·2.

In this group selenium is the mean between sulphur and tellurium.

Group VII.—Nitrogen, 14 ; phosphorus, 31 ; arsenic, 75 ; osmium, 99·6 ; antimony, 120·3 ; bismuth, 213.

13. First seven of Newlands' grouping of elements in 1863.

From a modern perspective, he had only placed thallium incorrectly since it belongs instead with boron, aluminium, gallium, and indium. The element thallium had been discovered just one year previously by the Englishman William Crookes. The first person to place it correctly in the boron group was a co-discoverer of the periodic system, namely Julius Lothar Meyer working in Germany. Even the great Mendeleev misplaced thallium in his early periodic tables and, like Newlands, he placed the element among the alkali metals.

In his first paper on classification of the elements, Newlands makes the following remark about the alkali metal group:

> The relation among the equivalents of this group may perhaps most simply be stated as follows, 1 of lithium (7) + 1 of potassium (39) = 2 of sodium.

This is, of course, nothing but a rediscovery of the triad relationship among these elements since,

Li	7	
Na	23	$2Na(23) = 7 + 39$
K	39	

In 1864, Newlands began publishing a series of articles in which he groped his way to a better periodic system and to what he later called the 'law of octaves', that is, the idea that elements repeat after moving through a sequence of eight. In 1865, he included sixty-five elements in his system and used ordinal numbers rather than atomic weights as a means of ordering the elements in a sequence of increasing weights. Now he began to write quite confidently of a new law, whereas De Chancourtois had only briefly considered this possibility but had rejected it.

Partly because of this choice of a musical analogy in discussing 'octaves' and because Newlands was not an academic chemist, his idea was ridiculed when he presented it verbally to the Royal

Society of Chemistry in 1866 (Figure 14). One member of this austere audience asked Newlands whether he had considered ordering the elements in alphabetical order. Newlands' paper was not published in the proceedings of the society, although he published further articles in a few other chemical journals. The time for the acceptance of the idea had still not come even though it was starting to occur to some chemists. But Newlands soldiered on and responded to his critics, while publishing a number of other periodic tables.

Odling

Another chemist to publish an early periodic table was William Odling, who, unlike Newlands, was a leading academic chemist. Odling attended the Karlsruhe conference and became a champion of Cannizzaro's views in Britain. He also held major appointments such as a professorship of chemistry at Oxford and the directorship of the Royal Institution in Albemarle Street in London. Odling independently published his own version of the periodic table by arranging the elements in order of increasing atomic weight, as Newlands had done, and by showing similar elements in horizontal rows (Figure 15).

In a paper written in 1864, Odling states that:

> Upon arranging the atomic weights or proportional numbers of the sixty or so recognized elements in order of their several magnitudes, we observe a marked continuity in the resulting arithmetical series.

This is followed by the further statement:

> With what ease this purely arithmetical seriation may be made to accord with a horizontal arrangement of the elements according to their usually received groupings, is shown in the following table, in the first three columns of which the numerical sequence is perfect, while in the other two the irregularities are but few and trivial.

No.	No.	No.	No.	No.	No.	No.	No.
H 1	F 8	Cl 15	Co & Ni 22	Br & Ni 22	Pd 36	I 42	Pt & Ir 50
Li 2	Na 9	K 16	Cu 23	Rb 30	Ag 37	Cs 44	Os 51
G 3	Mg 10	Ca 17	Zn 24	Sr 31	Cd 38	Ba & V 45	Hg 52
Bo 4	Al 11	Cr 19	Y 25	Ce & La 33	U 40	Ta 46	Tl 53
C 5	Si 12	Ti 18	In 26	Zr 32	Sn 39	W 47	Pb 54
N 6	P 23	Mn 20	As 27	Di & Mo 34	Sb 41	Nb 48	Bi 55
O 7	S 14	Fe 21	Se 28	Ro & Ru 35	Te 43	Au 49	Th 56

14. Newlands' table illustrating the law of octaves as presented to the Chemical Society in 1866.

			Bo 104	Pt 197
			Bu 104	Ir 197
			Pd 106.5	Os 199
H 1	"	"	Ag 108	Au 196.5
"	"	Zn 65	Cd 112	Hg 200
L 7	"	"	"	Tl 203
G 9	"	"	"	Pb 207
B 11	Al 27.5	"	U 120	"
C 12	Si 28	"	Sn 118	"
N 14	P 31	As 75	Sb 122	Bi 210
O 16	S 32	Se 79.5	Te 129	"
F 19	Cl 35.5	Br 80	I 127	"
Na 23	K 39	Bb 85	Cs 133	"
Mg 24	Ca 40	Sr 87.5	Ba 137	"
	Ti 50	Zr 89.5	Ta 138	Th 231.5
	"	Ce 92	"	
	Cr 52.5	Mo 96	V 137	
	Mn 55		W 184	
	Fe 56			
	Co 59			
	Ni 59			
	Cu 63.5			

15. Odling's third table of differences.

It is not clear why Odling's discovery was not accepted, since he did not lack academic credentials. It seems to have been due to the fact that Odling himself lacked enthusiasm for the idea of chemical periodicity and was reluctant to believe that it might represent a law of nature.

Hinrichs

In the United States, a newly arrived Danish immigrant, Gustavus Hinrichs, was busy developing his own system of classification of the elements, which he published in a striking radial format. However, Hinrichs' writing tended to be shrouded in layers of mysterious allusions to Greek mythology and other eccentric excesses, not to mention the fact that he managed to alienate himself from colleagues and the mainstream community of chemists.

Hinrichs was born in 1836 in Holstein, then a part of Denmark but later becoming a German province. He published his first book at the age of twenty, while attending the University of Copenhagen. He emigrated to the United States in 1861 to escape political persecution, and after a year of teaching in a high school was appointed head of modern languages at the University of Iowa. A mere one year later, he became Professor of Natural Philosophy, Chemistry, and Modern Languages. He is also credited with founding the first meteorological station in the USA, and acting as its director for fourteen years. One of the few detailed accounts of Hinrichs' life and work was published by Karl Zapffe, who says:

> It is not necessary to read far into Hinrichs' numerous publications to recognize the marks of an egocentric zeal which defaced many of his contributions with an untrustworthy eccentricity. Only at this late date does it become possible to separate those inspirations which were real—and which swept him off his feet—from background material which he captured in the course of his own learning. Whatever the source, Hinrichs usually dressed it with multilingual ostentation, and to such a point of disguise that he even came to regard Greek philosophy as his own.

Hinrichs' wide range of interests extended to astronomy. Like many authors before him, as far back as Plato, Hinrichs noticed some numerical regularities regarding the sizes of the planetary orbits. In an article published in 1864, he showed the table in Figure 16, which he proceeded to interpret.

Hinrichs expressed the differences in these distances by the formula $2^x \times n$, in which n is the difference in the distances of Venus and Mercury from the Sun, or twenty units. Depending on the value of x, the formula therefore gives the following distances:

$$2^0 \times 20 = 20$$
$$2^1 \times 20 = 40$$
$$2^2 \times 20 = 80$$
$$2^3 \times 20 = 160$$
$$2^4 \times 20 = 320$$

etc.

Distance to the Sun	
Mercury	60
Venus	80
Earth	120
Mars	200
Asteroid	360
Jupiter	680
Saturn	1320
Uranus	2600
Neptune	5160

16. Hinrichs' table of planetary distances (1864).

A few years previously, in 1859, the Germans Gustav Kirchhoff and Robert Bunsen had discovered that each element could be made to emit light, which could then be dispersed with a glass prism and analysed quantitatively. What they also discovered was that every single element gave a unique spectrum consisting of a set of specific spectral lines, which they set about measuring and publishing in elaborate tables. Some authors suggested that these spectral lines might provide information about the various elements that had produced them, but these suggestions met with strenuous criticism from one of their discoverers, Bunsen. Indeed, Bunsen remained quite opposed to the idea of studying spectra in order to gain information on atoms or to classify them in some way.

Hinrichs, however, had no hesitation in connecting spectra with the atoms of the elements. In particular, he became interested in the fact that, with any particular element, the frequencies of its spectral lines always seemed to be whole-number multiples of the smallest difference. For example, in the case of calcium, a ratio of 1:2:4 had been observed among its spectral frequencies. Hinrichs' interpretation of this fact was bold and elegant: if the sizes of planetary orbits produce a regular series of whole numbers, as mentioned earlier, and if the ratios among spectral line differences also produce whole-number ratios, the cause of the latter might lie in the size ratios among the atomic dimensions of the various elements.

The eleven 'spokes' radiating from the centre of Hinrichs' wheel-like system consist of three predominantly non-metal groups and eight groups containing metals (Figure 17). From a modern perspective, the non-metal groups appear to be incorrectly ordered, in that the sequence is groups 16, 15, and then 17 when proceeding from left to right on the top of the spiral. The group containing carbon and silicon is classed with

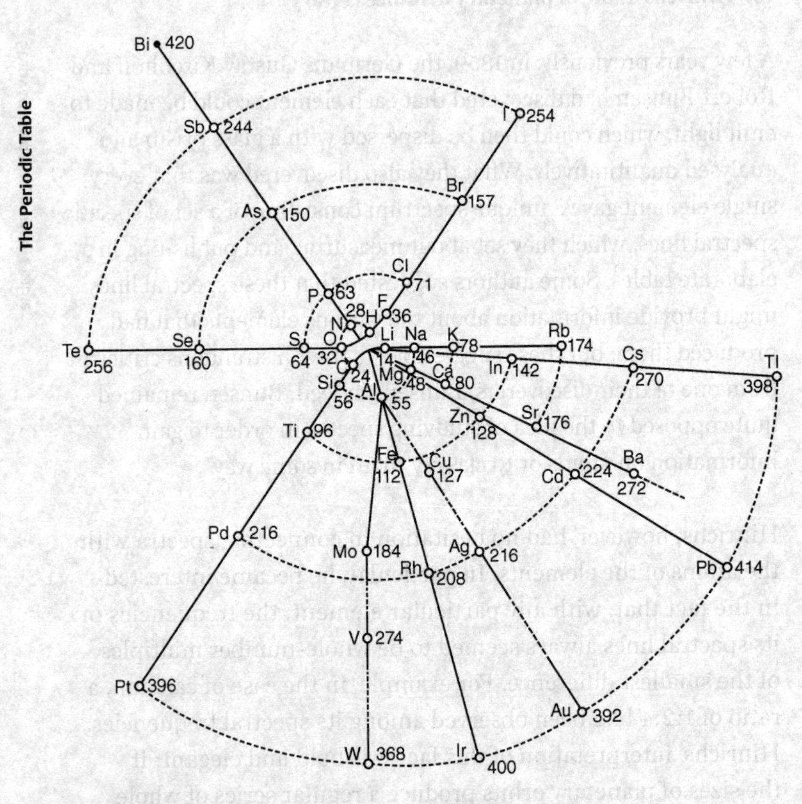

17. Hinrichs' periodic system.

the metallic groups by Hinrichs, presumably because it also includes the metals nickel, palladium, and platinum. In the modern table, these three metals are indeed grouped together, but not in the same group as carbon and silicon, which belong with germanium, tin, and lead in group 14.

Overall, however, Hinrichs' periodic system is rather successful in grouping together many important elements. One of its main advantages is the clarity of its groupings, as compared to the more elaborate but less successful periodic systems of Newlands in 1864 and 1865 for example. Hinrichs possessed a deep knowledge of chemistry, as well as a proficiency in mineralogy. He was perhaps the most interdisciplinary of all the discoverers of the periodic system. The fact that Hinrichs arrived at his system from such a different direction to the others might be taken to lend the periodic system itself independent support.

In an article published in *The Pharmacists* in 1869, Hinrichs discusses previous unsuccessful attempts to classify the elements, but in doing so fails to mention any of his co-discoverers such as De Chancourtois, Newlands, Odling, Lothar Meyer, or Mendeleev. Hinrichs characteristically appears to have completely ignored all other attempts to base the classification of the elements directly on atomic weights, though one can assume that he may have been aware of them given his knowledge of foreign languages.

Lothar Meyer

The first periodic system to have something of an impact on the scientific world was due to the German chemist Julius Lothar Meyer working in Jena. But Lothar Meyer is generally regarded as the runner-up to Mendeleev in terms of who discovered the periodic system proper. This assessment is generally true, but a number of aspects of his work suggest that he might well be regarded as the co-discoverer instead of an also-ran.

Like Mendeleev, Lothar Meyer attended the Karlsruhe conference as a young man. He seems to have been very impressed by the ideas that Cannizzaro presented at this meeting and was soon editing the German edition of Cannizzaro's works. In 1862, just two years after Karlsruhe, Lothar Meyer devised two partial periodic tables, one consisting of twenty-eight elements arranged in order of increasing atomic weight in which the elements were grouped into vertical columns according to their chemical valences (Figure 18).

In 1864, he published an influential book on theoretical chemistry in which he included his two tables. The second table featured twenty-two elements and was partly arranged in increasing atomic weight order.

Lothar Meyer's approach was one of a theoretical or physical chemist. He placed more emphasis on such quantities as densities, atomic volumes, and melting points of the elements than on their chemical properties. Contrary to the popular account, Lothar Meyer *did* leave some gaps in his periodic tables and even ventured to predict some properties of the elements that might eventually fill these spaces. One such prediction was for an element that was isolated in 1886 and was named germanium. Unlike Mendeleev, Lothar Meyer believed in the essential unity of all matter and was a supporter of Prout's hypothesis of the composite nature of the elements.

In 1868, he produced an expanded periodic system for the second edition of his textbook which contained fifty-three known elements (Figure 19). Unfortunately, this table was misplaced by the publisher. It did not appear in the new edition of his book nor in any journal articles. Even Lothar Meyer himself seems to have forgotten the existence of this table, since he never mentioned it after a priority dispute later developed with Mendeleev. Had this table been brought to light at that time, it is not clear that Mendeleev's claim to precedence would have been afforded quite the weight that it is given today.

	4 werthig	3 werthig	2 werthig	1 werthig	1 werthig	2 werthig
	--	--	--	--	Li = 7.03	(Be = 9.3?)
Differenz =	--	--	--	--	16.02	(14.7)
	C = 12.0	N = 14.04	O = 16.00	Fl = 19.0	Na = 23.05	Mg = 24.0
Differenz =	16.5	16.96	16.07	16.46	16.08	16.0
	Si = 28.5	P = 31.0	S = 32.07	Cl = 3S.46	K = 39.13	Ca = 40.0
Differenz =	$\frac{89.1}{2} = 44.55$	44.0	46.7	44.51	46.3	47.6
	--	As = 75.0	Se = 78.8	Br = 79.97	Rh = 85.4	Sr = 87.6
Differenz =	$\frac{89.1}{2} = 44.55$	45.6	49.5	46.8	47.6	49.5
	Sn = 117.6	Sb = 120.6	Te = 128.3	I = 126.8	Cs = 133.0	Ba = 137.1
Differenz =	89.4 = 2 × 44.7	87.4 = 2 × 43.7	--	--	(71 = 2 × 35.5)	--
	Pb = 207.0	Bi = 208.0	--	--	(Tl = 204?)	--

18. Lothar Meyer's periodic system of 1862.

1	2	3	4	5	6	7	8
		Al = 27.3 $\frac{28\ 7}{2} = 14.8$	Al. = 27.3				C = 12.00 16.5 Si = 28.5 $\frac{89.1}{2} = 44.55$
Cr = 52.6	Mn = 55.1 49.2 Ru = 104.3 92.8 = 2.46.4 Pt = 197.1	Fe = 56.0 48.9 Rh=103.4 92.8=2.46.4 Ir = 197.1	Co = 58.7 47.8 Pd = 106.0 93 = 2.465 Os = 199.	Ni = 58.7	Cu = 63.5 44.4 Ag = 107.9 88.8 = 2.44.4 Au = 196.7	Zn = 65.0 46.9 Cd = 111.9 88.3 = 2.44.5 Hg = 200.2	$\frac{89.1}{2} = 44.55$ Su = 117.6 89.4 = 2.41.7 Pb = 207.0
9	**10**	**11**	**12**	**13**	**14**	**15**	
N = 14.4 16.96 P = 31.0 44.0 AS = 75.0 45.6 Sb = 120.6 87.4 = 2.43.7 Bi = 208.0	O = 16.00 16.07 S = 32.07 46.7 Se = 78.8 49.5 Te = 128.3	F = 19.0 16.46 Cl = 35.46 44.5 Br = 79.9 46.8 I = 126.8	Li = 7.03 16.02 Na = 23.05 16.08 K = 39.13 46.3 Rb = 85.4 47.6 Cs = 133.0 71 = 2.35.5 Te = 204.0	Be = 9.3 14.7 Mg = 24.0 16.0 Ca = 40.0 47.6 Sr = 87.6 49.5 Ba = 137.1	Ti = 48 42.0 Zr = 90.0 47.6 Ta = 137.6	Mo. = 92.0 45.0 Vd = 137.0 47.0 W = 184.0	

19. Lothar Meyer's periodic system of 1868.

Lothar Meyer's lost table has great merit for its inclusion of so many elements and for certain placements which Mendeleev's famous table of the same year failed to achieve. The lost table was eventually published, after Lothar Meyer's death, in 1895 and far too late to have any impact on the question of who had arrived at the first fully mature periodic system.

During the course of the rather public dispute, Mendeleev took a more forceful approach, claiming that the credit belonged to himself alone because he had gone beyond discovering the periodic system by making a number of successful predictions. Lothar Meyer appears to have adopted something of a defeatist attitude by even admitting that he had lacked the courage to make predictions.

Chapter 5
The Russian genius: Mendeleev

Dmitri Ivanovich Mendeleev is by far the most famous Russian scientist of the modern era. Not only did he discover the periodic system, but he recognized that it pointed to a deep law of nature, the periodic law. He also spent many years drawing out the full consequences of this law, most notably by predicting the existence and properties of many new elements. In addition, he corrected the atomic weights of some already known elements and successfully changed the position of some other elements in the periodic table.

But perhaps most important of all, Mendeleev made the periodic table his own by pursuing its study and development during several periods of his life, even though he worked in a number of other very diverse fields. By contrast, most of his precursors or co-discoverers failed to follow up their initial discoveries. As a result, the name of Mendeleev is inextricably linked with the periodic table in much the same way that evolution by natural selection and relativity theory are linked with Darwin and Einstein respectively. Given Mendeleev's supreme importance in the story of the periodic table, this entire chapter will be devoted to his scientific work as well as his early development.

When Mendeleev was very young, his father, who owned a glass factory, went blind and died soon afterwards. Dmitri, the youngest of fourteen siblings, was brought up by his doting mother, who was determined to provide the best possible education for him. This devotion included travelling for hundreds of miles with the young man in a failed attempt to enrol him into Moscow University. The reason for Mendeleev's being rejected was apparently his Siberian origin and the fact that the university admitted only Russians. Undeterred, Mendeleev's mother succeeded in getting the young man into the Main Pedagogical Institute of St Petersburg, where he began to study chemistry, physics, biology, and of course pedagogy, the last of which was to have a telling effect on his discovering the mature periodic system. Sadly, Mendeleev's mother died very soon after he had entered this institution, and he was left to fend for himself.

After completing his undergraduate education, Mendeleev spent some time in France, and later Germany, where he was formally attached to the laboratory of Robert Bunsen, although he preferred to stay at home and conduct his private experiments on gases. It was during this period in Germany that Mendeleev attended the Karlsruhe conference in 1860, not because he was a prominent chemist but more because he happened to be at the right place at the right time. Although Mendeleev quickly grasped the value of Cannizzaro's ideas, as Lothar Meyer did at this conference, Mendeleev's conversion to using Cannizzaro's atomic weights appears to have taken considerably more time than it did for Lothar Meyer.

In 1861, Mendeleev began to show real promise when he published a textbook on organic chemistry which won him the coveted Demidov Prize in Russia. In 1865, he defended his doctoral thesis on the interaction between alcohol and water, and began to work on a book about inorganic chemistry in order to improve the teaching of chemistry. In the first volume of this new book, he treated the more common elements in no particular order. By 1868, he had completed this volume and had begun to consider how he should make the transition to the remaining elements in a second volume.

The actual discovery

Although Mendeleev had been thinking about elements, atomic weights, and classification for about ten years, he appears to have had his eureka moment, or perhaps 'eureka day', on 17 February 1869. On this day, he cancelled his trip to visit a cheese factory as a consultant and decided to work on what was to become his famous brainchild—the periodic table.

First, he listed the symbols for a handful of elements in two rows on the back of the invitation to the cheese factory:

Na	K	Rb	Cs
Be	Mg	Zn	Cd

Then he produced a slightly larger array of sixteen elements:

F	Cl	Br	I				
Na	K	Rb	Cs			Cu	Ag
Mg	Ca	Sr	Ba	Zn	Cd		

By that evening, Mendeleev had sketched out an entire periodic table which included sixty-three known elements. Moreover, it included several gaps for as yet unknown elements and even the predicted atomic weights for some of these elements. Two hundred copies of this first table were printed and sent to chemists all over Europe. On 6 March of the same year, the discovery was announced by one of Mendeleev's colleagues at a meeting of the Russian Chemical Society. Within a month, an article had appeared in the journal of this newly formed society and another longer article appeared in Germany.

Many popular books and documentary programmes about Mendeleev claim that he arrived at his periodic table in the course of a dream or while laying out cards for each of the elements as though playing the game of patience/solitaire. The second of these tales, especially, is now regarded as being apocryphal by Mendeleev biographers such as science historian Michael Gordin.

To turn to Mendeleev's scientific approach, it seems that he differed considerably from his rival Lothar Meyer in not believing in the unity of all matter and similarly not supporting Prout's hypothesis for the composite nature of all elements. Mendeleev also took care to distance himself from the notion of triads of elements. For example, he proposed that the element fluorine should be grouped together with chlorine, bromine, and iodine even though this would imply going beyond a triad to form a group of at least four elements.

Whereas Lothar Meyer had concentrated on physical principles, and mainly on the physical properties of the elements, Mendeleev was very familiar with their chemical properties. On the other hand, when it came to deciding upon the most important criterion for classifying the elements, Mendeleev insisted that atomic weight ordering would tolerate no exceptions. Of course, many of Mendeleev's precursors, such as De Chancourtois, Newlands, Odling, and Lothar Meyer, had recognized the importance of atomic weight ordering to varying degrees. But Mendeleev also reached a deeper philosophical understanding of atomic weights and the nature of the elements which allowed him to move into uncharted territory regarding as yet unknown elements.

The nature of elements

There was a long-standing puzzle in chemistry. When sodium and chlorine combine, they give rise to a completely new substance, sodium chloride, in which the combining elements do not seem to survive, at least from a visual perspective. This is the phenomenon of chemical bonding, or chemical combination, which differs markedly from the physical mixing of, say, powdered sulphur and iron filings.

One of the questions, in the case of chemical combination, is to understand how, if at all, the combining elements survive in the compound. This issue is further complicated in several languages, such as English, in which we use the word 'element' to refer to the combined substance such as chlorine when it is present in sodium chloride. Now what underlies both the uncombined green gas chlorine and combined chlorine is sometimes also called 'element'. We now have three senses of the same central chemical term to describe the substances that the periodic table is supposed to classify.

The third sense of 'element', as introduced above, has been variously called metaphysical element, abstract element, transcendental element, and more recently 'element as basic substance'. This is the element as an abstract bearer of properties but lacking such phenomenal properties as the green colour of chlorine. Green chlorine, meanwhile, is said to be the element as a 'simple substance'.

When Antoine Lavoisier revolutionized chemistry at the end of the 18th century, one of his contributions was to concentrate attention onto the element as a simple substance, that is, the element in isolated form. This was intended to improve chemistry by ridding it of excess metaphysical baggage and it was indeed a great step forward. Now an 'element' was to be regarded primarily as the final step in the separation of the components in any compound. Whether Lavoisier intended to banish the more abstract and philosophical sense of 'element' is a moot point, but certainly this sense began to take a back-seat to the sense of an element that could be isolated.

However, the more abstract sense was not completely forgotten, and Mendeleev was one of the chemists who not only understood it but proposed to elevate its status. In fact, he repeatedly claimed that the periodic system was primarily a classification of this more abstract sense of the term 'element' and not necessarily the more concrete element that could be isolated.

The reason why I have carefully developed this issue is that, when he was armed with this notion, Mendeleev could take a deeper view of elements than chemists who restricted themselves to elements in isolated form. It afforded Mendeleev the possibility of going beyond appearances. If a particular element did not appear to fit within a particular group, Mendeleev could draw upon the deeper sense of the term 'element', and thus to some extent he could ignore the apparent properties of the element as an isolated or simple substance.

Mendeleev's predictions

One of his greatest triumphs, and perhaps the one that he is best remembered for, is Mendeleev's correct prediction of the existence of several new elements. In addition, he corrected the atomic weights of some elements as well as relocating other elements to new positions within the periodic table. As I suggested in the previous section, such far-sightedness may have been due to his having a greater philosophical understanding of the nature of the elements than his competitors. By concentrating on the more abstract concept of elements as basic substances, Mendeleev was able to surmount apparent obstacles that arose from taking the properties of the isolated elements at face value.

Although Mendeleev placed the greatest importance on the atomic weights of the elements he also considered chemical and physical properties and family resemblances among them. Whereas Lothar Meyer concentrated on physical properties, Mendeleev paid more attention to the chemistry of the elements. Another criterion he used was that each element should occupy a single place in the periodic table, although he was willing to violate this notion when it came to what he called group VIII (Figure 5). His more important criterion, however, was the ordering of the elements according to increasing atomic weights. In one or two cases, he appeared to even violate this principle, although a closer inspection shows that this was not in fact the case.

The case of tellurium and iodine is one of only four pair reversals in the periodic system and the best known among them. These are elements which appear in the reverse order than they would according to increasing values of atomic weight (Te = 127.6, I = 126.9, and yet Te appears before I). Many historical accounts make a point of recounting how clever Mendeleev was to reverse the positions of these elements, thus putting chemical properties over and above the ordering according to atomic weight. Such a claim is incorrect in several respects. First of all, Mendeleev was by no means the first chemist to make this particular reversal. Odling, Newlands, and Lothar Meyer all published tables in which the positions of tellurium and iodine had been reversed, well before the appearance of Mendeleev's articles. Second, Mendeleev was not in fact placing a greater emphasis on chemical properties than on atomic weight ordering.

Mendeleev held to his criterion of ordering according to increasing atomic weight and repeatedly stated that this principle would tolerate no exceptions. His thinking regarding tellurium and iodine was rather that the atomic weights for one or both of these elements had been incorrectly determined, and that future work would reveal that, even on the basis of atomic weight ordering, tellurium should be placed before iodine. On this point, as in many instances that tend to go unreported, Mendeleev was wrong.

At the time when Mendeleev proposed his first periodic systems, the atomic weights for tellurium and iodine were thought to be 128 and 127 respectively. Mendeleev's belief that atomic weight was the fundamental ordering principle meant that he had no choice but to question the accuracy of these two values. This was because it was clear that in terms of chemical similarities, tellurium should be grouped with the elements in group VI and iodine with those in group VII, or, in other words, that this pair of elements should be 'reversed'. Mendeleev continued to question the reliability of these atomic weights until the end of his life.

Initially, he doubted the atomic weight of tellurium while believing that that of iodine was essentially correct. Mendeleev began to list tellurium as having an atomic weight of 125 in some of his subsequent periodic tables. At one time, he asserted that the commonly reported value of 128 was the result of measurements having been made on a mixture of tellurium and a new element he called eka-tellurium. Motivated by these claims, the Czech chemist Bohuslav Brauner began a series of experiments in the early 1880s aimed at the redetermination of the atomic weight of tellurium. By 1883, he reported that the value for tellurium should be 125. Mendeleev was sent a telegram of congratulations by other participants present at the meeting at which Brauner had made this announcement. In 1889, Brauner obtained new results that seemed to further strengthen the earlier finding of Te = 125.

But in 1895, everything changed as Brauner himself started reporting a new value for tellurium that was greater than that of iodine, thus returning matters to their initial starting point. Mendeleev's response was now to begin to question the accuracy of the accepted atomic weight for iodine instead of tellurium. This time, he requested a redetermination for the atomic weight of iodine and hoped that its value would turn out to be higher. In some of his later periodic tables, Mendeleev even listed tellurium and iodine as both having atomic weights of 127. The problem would not be resolved until work by Henry Moseley in 1913 and 1914, who showed that the elements should be ordered according to atomic number rather than atomic weight. While tellurium has a higher atomic weight than iodine it has a lower atomic number, and this is why it should be placed before iodine, in full agreement with its chemical behaviour.

Although Mendeleev's predictions might seem quite miraculous, they were in fact based on careful interpolation between the properties of known elements flanking the unknown ones. He began to make predictions in his very first publication on the periodic system in 1869, although he published a more detailed account of his forecasts in a long paper of 1871. He began by focusing on two gaps, one below aluminium and the other below silicon in his periodic table, which he provisionally called eka-aluminium and eka-boron, using the Sanskrit prefix meaning 'one' or 'one-like'. In his paper of 1869, he wrote:

> We must expect the discovery of yet unknown elements, e.g. elements analogous to Al and Si, with atomic weights 65–75.

In the autumn of 1870, he began to speculate about a third element, which would lie below boron in the table. He listed the atomic volumes of the three elements as

eka-boron	eka-aluminium	eka-silicon
15	11.5	13

In 1871, he predicted their atomic weights to be

eka-boron	eka-aluminium	eka-silicon
44	68	72

and he gave a detailed set of predictions on various chemical and physical properties of all three elements.

It took six years before the first of these predicted elements, later called gallium, was isolated. With a few very minor exceptions,

Property	Eka-silicon 1871 prediction	Germanium discovered 1886
Relative atomic mass	72	72.32
Specific gravity	5.5	5.47
Specific heat	0.073	0.076
Atomic volume	13 cm^3	13.22 cm^3
Colour	Dark grey	Greyish white
Specific gravity of dioxide	4.7	4.703
Boiling point of tetrachloride	100°C	86°C
Specific gravity of tetrachloride	1.9	1.887
Boiling point of tetra ethyl deriv.	160°C	160°C

20. The predicted and observed properties of eka-silicon (germanium).

Mendeleev's predictions were almost exactly correct. The accuracy of Mendeleev's predictions can also be clearly seen in the case of the element he called eka-silicon, later called germanium, after it had been isolated by the German chemist Clemens Winkler. It would seem that Mendeleev's only failure lies in the specific boiling point of the tetrachloride of germanium (Figure 20).

Mendeleev's failed predictions

But not all of Mendeleev's predictions were so dramatically successful, a feature that seems to be omitted from most popular accounts of the history of the periodic table. As Figure 21 shows, he was unsuccessful in as many as eight out of his eighteen published predictions, although perhaps not all of these predictions should be given the same weight. This is because some of the elements involved the rare earths, which resemble each other very closely and which posed a major challenge to the periodic table for many years to come.

Element as given by Mendeleev.	Predicted A.W.	Measured A.W.	Eventual name
coronium	0.4	not found	not found
ether	0.17	not found	not found
eka-boron	44	44.6	scandium
eka-cerium	54	not found	not found
eka-aluminium	68	69.2	gallium
eka-silicon	72	72.0	germanium
eka-manganese	100	99	technetium (1939)
eka-molybdenum	140	not found	not found
eka-niobium	146	not found	not found
eka-cadmium	155	not found	not found
eka-iodine	170	not found	not found
eka-caesium	175	not found	not found
tri-manganese	190	186	rhenium (1925)
dvi-tellurium	212	210	polonium (1898)
dvi-caesium	220	223	francium (1939)
eka-tantalum	235	231	protactinium (1917)

21. Mendeleev's predictions, successful and otherwise.

In addition, Mendeleev's failed predictions raise another philosophical issue. For a long time, historians and philosophers of science have debated whether successful predictions made in the course of scientific advances should count more, or less, than the successful accommodation of already known data. Of course, there is no disputing the fact that successful predictions carry a greater psychological impact since they almost imply that the scientist in question can foretell the future. But successful accommodation, or explanation of already known data, is no mean feat either, especially since there is usually more already known information to incorporate into a new scientific theory. This was especially the case with Mendeleev and the periodic table, since he had to successfully accommodate as many as sixty-three known elements into a fully coherent system.

At the time of the discovery of the periodic table, the Nobel Prizes had not yet been instituted. One of the highest accolades in chemistry was the Davy Medal, awarded by the Royal Society of Chemistry in Britain and named after the chemist Humphry Davy. In 1882, the Davy Medal was jointly awarded to Lothar Meyer and Mendeleev. This fact seems to suggest that the chemists making the award were not overly impressed by Mendeleev's successful predictions, given that they were willing to recognize Lothar Meyer, who had not made any predictions to speak of. Moreover, the detailed citation that accompanied the award to Lothar Meyer and Mendeleev made absolutely no mention of Mendeleev's successful predictions. It would appear that at least this group of eminent British chemists were not swayed by the psychological impact of successful predictions over the ability of an individual to successfully accommodate the known elements.

The noble gases

The discovery of the noble gases at the end of the 19th century represented an interesting challenge to the periodic system for a number of reasons. First of all, in spite of Mendeleev's dramatic predictions of many other elements, he completely failed to predict this entire group of elements (He, Ne, Ar, Kr, Xe, Rn).

The first of them to be isolated was argon, in 1894, at University College London. Unlike many previously discussed elements, a number of factors conspired to render the accommodation of this element something of a Herculean task. It was not until six years later that the inert gases took their place as an eighth group between the halogens and the alkali metals.

But let us return to the first noble gas to be isolated, namely argon. It was obtained in small amounts, by Lord Rayleigh and William Ramsey who were working with the gas nitrogen. The all-important atomic weight of argon that was required if it were to be placed in the periodic table was not easily obtained. This was due to the fact that the atomicity of argon was itself not easy to determine. Most measurements pointed to its being monoatomic, but all the other known gases at the time were diatomic (H_2, N_2, O_2, F_2, Cl_2). If argon were indeed monoatomic its atomic weight would be approximately 40, which rendered its accommodation into the periodic table somewhat problematic, since there was no gap at this value of atomic weight. The element calcium has an atomic weight of about 40, followed by scandium, one of Mendeleev's successfully predicted elements, with an atomic weight of 44. This seemed to leave no space for a new element with an atomic weight of 40 (Figure 5).

There was a rather large gap between chlorine (35.5) and potassium (39), but placing argon between these two elements would have produced a rather obtrusive pair reversal. It should be recalled that at this time only one significant such pair reversal existed, involving the elements tellurium and iodine, and this behaviour was regarded as being highly anomalous. Mendeleev had concluded that the Te–I pair reversal was due to incorrect atomic weight determinations on either tellurium or iodine or perhaps both elements.

Yet another unusual aspect of the element argon was its seemingly complete chemical inertness, which meant that its compounds could not be studied, for the simple reason that it did not form any. Some researchers took the inertness of the gas to mean that it was not a genuine chemical element. If this were true, it would provide an easy way out of the quandary as it would not need to be placed anywhere.

But many others persisted in attempting to place the element into the periodic table. The accommodation of argon was the focus of a general meeting of the Royal Society in 1885, attended by the leading chemists and physicists of the day. The discoverers of argon, Rayleigh and Ramsey, argued that the element was probably monoatomic but admitted that they could not be sure. Nor could they be sure that the gas in question was not a mixture, which might imply that the atomic weight was not in fact 40. William Crookes presented some evidence in favour of sharp boiling and melting points of argon, thus pointing to a single element rather than a mixture. Henry Armstrong, a leading chemist, argued that argon might behave like nitrogen, in that it would form an inert diatomic molecule even though its individual atoms might be highly reactive.

A physicist, Arthur William Rücker, argued that an atomic weight of approximately 40 might be correct and that if this element could not be housed in the periodic table, then it was the periodic table itself that was at fault. This comment is interesting because it shows that even sixteen years after the publication of Mendeleev's periodic table, and even after his three famous predictions had been fulfilled, not everyone was convinced of the validity of the periodic system.

The findings of the Royal Society meeting were therefore somewhat inconclusive about the fate of the new element argon, or whether indeed it should count as a new element. Mendeleev himself, who had not attended this meeting, published an article in the London-based *Nature* magazine in which he concluded that argon was in fact tri-atomic and that it consisted of three atoms of nitrogen. He based this conclusion on the fact that the assumed atomic weight of 40 divided by three gives approximately 13.3, which is not so far removed from 14, the atomic weight of nitrogen. In addition, argon had been discovered in the course of experiments on nitrogen gas, which rendered the tri-atomic idea more plausible.

The issue was finally resolved in the year 1900. At a conference in Berlin, Ramsey, one of the co-discoverers of the new gas, informed Mendeleev that the new group, which had by then been augmented with helium, neon, krypton, and xenon, could be elegantly accommodated in an eighth column between the halogens and the alkali metals. Argon, the first of these new elements to be discovered, had been particularly troublesome because it represented a new case of a genuine pair reversal. It has an atomic weight of about 40 and yet appears before potassium with an atomic weight of about 39. Mendeleev now readily accepted this proposal and later wrote:

> This was extremely important for him [Ramsey] as an affirmation of the position of the newly discovered elements, and for me as a glorious confirmation of the general applicability of the periodic law.

Far from threatening the periodic table, the discovery of the inert gases and their successful accommodation into the periodic table only served to underline the great power and generality of Mendeleev's periodic system.

Chapter 6
Physics invades the periodic table

Although John Dalton had reintroduced the notion of atoms to science, many debates followed among chemists, most of whom refused to accept that atoms existed literally. One of these sceptical chemists was Mendeleev, but, as we have seen in previous chapters, this does not seem to have prevented him from publishing the most successful periodic system of all those proposed at the time. Following the work of physicists like Einstein, the atom's reality became more and more firmly established, starting at the turn of the 20th century. Einstein's 1905 paper on Brownian motion, using statistical methods, provided conclusive theoretical justification for the existence of atoms but lacked experimental support. The latter was soon provided by the French experimental physicist Jean Perrin.

This change was accompanied by many lines of research aimed at exploring the structure of the atom, and developments which were to have a major influence on attempts to understand the periodic system theoretically. In this chapter, we consider some of this atomic research, as well as several other key discoveries in 20th-century physics that contributed to what I will call the invasion of the periodic table by physics.

The discovery of the electron, the first subatomic particle and the first hint that the atom had a substructure, came in 1897 at the hands of the legendary J. J. Thomson at the Cavendish Laboratory in Cambridge. A little earlier, in 1895, Wilhelm Konrad Röntgen had discovered X-rays in Würzburg, Germany. These new rays would soon be put to very good use by Henry Moseley, a young physicist working first in Manchester and, for the remainder of his short scientific life, in Oxford.

Just a year after Röntgen had described his X-rays, Henri Becquerel in Paris discovered an enormously important phenomenon called radioactivity, whereby certain atoms were breaking up spontaneously while emitting a number of different new kinds of rays. The term 'radioactivity' was actually coined by the Polish-born Marie Sklodowska (later Curie) (Figure 22) who, with her husband Pierre Curie, took up the work on this dangerous new phenomenon and quickly discovered a couple of new elements that they called polonium and radium.

By studying how atoms break up while undergoing radioactive decay, it became possible to probe the components of the atom more effectively, as well as the laws that govern how atoms transform themselves into other atoms. So although the periodic table deals with distinct individuals or atoms of different elements, there seem to be some features that allow some atoms to be converted into others under the right conditions. For example, the loss of an alpha particle, consisting of a helium nucleus with two protons and two neutrons, results in a lowering of the atomic number of an element by two units.

Another very influential physicist working at this time was Ernest Rutherford, a New Zealander who arrived in Cambridge to pursue a research fellowship, later spending periods of time at McGill University in Montreal and at Manchester University. He then returned to Cambridge to assume the directorship of the Cavendish Laboratory as the successor to J. J. Thomson. Rutherford's contributions to atomic physics were many and varied,

22. Marie Curie.

and included the discovery of the laws governing radioactive decay as well as being the first to 'split the atom'. He was also the first to achieve the 'transmutation' of elements into other new elements. In this way, Rutherford achieved an artificial analogue to the process of radioactivity, which similarly yielded atoms of a completely different element and once again emphasized the essential unity of all forms of matter, another notion, incidentally, that Mendeleev had strenuously opposed during his lifetime.

Another discovery by Rutherford consisted of the nuclear model of the atom, a concept that is taken more or less for granted these days, namely the notion that the atom consists of a central nucleus surrounded by orbiting negatively charged electrons. His Cambridge predecessor, Thomson, had thought that the atom consisted of a sphere of positive charge within which the electrons circulated in the form of rings.

Nevertheless, Rutherford was not the first to suggest a nuclear model of the atom which resembled a miniature solar system. That distinction belongs to the French physicist Jean Perrin, who in 1900 proposed that negative electrons circulated around the positive nucleus like planets circulating around the Sun. In 1903, this astronomical analogy was given a new twist by Hantaro Nagaoka, from Japan, who proposed his Saturnian model in which the electrons now took the place of the famous rings around the planet Saturn. But neither Perrin nor Nagaoka could appeal to any solid experimental evidence to support their models of the atom, whereas Rutherford could.

Rutherford, along with his junior colleagues Geiger and Marsden, fired a stream of alpha particles at a thin metal foil made of gold and obtained a very surprising result. Whereas most of the alpha particles passed through the gold foil in a more or less unimpeded fashion, a significant number of them were deflected at very oblique angles. Rutherford's conclusion was that atoms of gold, or anything else for that matter, consist mostly of empty space except for a dense central nucleus. The fact that some of the alpha particles were unexpectedly bouncing back towards the incoming stream of alpha particles was thus evidence for the presence of a tiny central nucleus in every atom.

Nature therefore turned out to be more fluid than had previously been believed. Mendeleev, for example, had thought that elements were strictly individual. He could not accept the notion that elements could be converted into different ones. In fact, after the Curies began to report experiments that suggested the breaking up of atoms, Mendeleev travelled to Paris to see the evidence for himself, close to the end of his life. It is not clear whether he accepted this radical new notion even after his visit to the Curie laboratory.

X-rays

In 1895, the German physicist Röntgen made a momentous discovery at the age of fifty. Up to this point, his research output had been somewhat unexceptional. As the atomic physicist Emilio Segrè wrote a good deal later, 'By the beginning of 1895 Röntgen had written forty-eight papers now practically forgotten. With his forty-ninth he struck gold.'

Röntgen was in the process of exploring the action of an electric current in an evacuated glass tube called a Crookes tube. Röntgen noticed that an object, which was not part of his experiment, on the other side of his lab was glowing. It was a screen coated with barium platinocyanide. He quickly established that the glow was not due to the action of the electric current and deduced that some new form of rays might be produced inside his Crookes tube. Soon, Röntgen also discovered the property for which X-rays are best known. He found that he could produce an image of his hand that clearly showed the outlines of just his bones. A powerful new technique that was to provide numerous medical applications had come to light. After working secretly for seven weeks, Röntgen was ready to announce his results to the Würzburg Physical-Medical Society, an interesting coincidence given the impact that his new rays were to have on these two fields that were still remote from each other at the time.

Some of Röntgen's original X-ray images were sent to Paris, where they reached Henri Becquerel, who became interested in examining the relationship between X-rays and the property of phosphorescence, whereby certain substances emit light on exposure to sunlight. In order to test this notion, Becquerel wrapped some crystals of a uranium salt in some thick paper and, because of a lack of sunlight, decided to put these materials away in a drawer for a few days. By another stroke of luck, Becquerel happened to place his wrapped crystals on top of an undeveloped photographic plate before closing the drawer and going about his business for a few more cloudy Parisian days.

On finally opening the drawer, he was amazed to find an image that had been created on the photographic plate by the uranium crystals even though no sunlight had struck them. This clearly suggested that the uranium salt was emitting its own rays, irrespective of the process of phosphorescence. Becquerel had discovered nothing less than radioactivity, a natural process in some materials whereby the spontaneous decay of the nuclei of atoms produces powerful, and in some cases dangerous emanations. It was Marie Curie who was to dub this phenomenon 'radioactivity' a few years later.

The supposed connection between these experiments and X-rays turned out to be incorrect. Becquerel failed to find any connection between X-rays and phosphorescence. In fact, X-rays had not even entered into these experiments, and yet he discovered a phenomenon that was to have enormous importance in more ways than one. First, radioactivity was an early and significant step into the exploration of matter and radiation, and second, it was to lead indirectly to the development of nuclear weapons.

Back to Rutherford

Around the year 1911, Rutherford reached the conclusion that the charge on the nucleus of the atom was approximately half of the weight of the atom concerned, or $Z \approx A/2$, after analysing the results of atomic scattering experiments. This conclusion was supported by the Oxford physicist Charles Barkla, who arrived at it via an altogether different route, using experiments with X-rays.

Meanwhile, a complete outsider to the field, the Dutch econometrician Anton van den Broek, was pondering over the possibility of modifying Mendeleev's periodic table. In 1907, he proposed a new table containing 120 elements, although many of these remained as empty spaces. A good number of empty spaces were occupied by some newly discovered substances whose elemental status was still in question. They included so-called thorium emanation, uranium X (an unknown decay product of uranium), Gd_2 (a decay product from gadolinium), and many other new species.

But the really novel feature of van den Broek's work was a proposal that all elements were composites of a particle that he named the 'alphon', consisting of half of a helium atom with a mass of two atomic weight units. In 1911, he published a further article in which he dropped any mention of alphons, but retained the idea of elements differing by two units of atomic weight. In a twenty-line letter to London's *Nature* magazine, he went a step further towards the concept of atomic number by writing, 'the number of possible elements is equal to the number of possible permanent charges'.

Van den Broek was thus suggesting that since the nuclear charge on an atom was half of its atomic weight, and the atomic weights of successive elements increased in step-wise fashion by two, then the nuclear charge defined the position of an element in the periodic table. In other words, each successive element in the periodic table would have a nuclear charge greater by one than the previous element.

A further article published in 1913 drew the attention of Niels Bohr, who cited van den Broek in his own famous trilogy paper of 1913 on the hydrogen atom and the electronic configurations of many-electron atoms. In the same year, van den Broek wrote another paper which also appeared in *Nature*, this time explicitly connecting the serial number on each atom with the charge on each atom. More significantly perhaps, he disconnected this serial number from atomic weight. This landmark publication was praised by many experts in the field, including Soddy and Rutherford, all of whom had failed to see the situation as clearly as the amateur van den Broek.

Moseley

Although an amateur had confounded the experts in arriving at the concept of atomic number, he did not quite complete the task of establishing this new quantity. The person who did complete it, and who is almost invariably given the credit for the discovery of atomic number, was the English physicist Henry Moseley, who died in World War I at the age of twenty-six. His fame rests on just two papers in which he confirmed experimentally that atomic number was a better ordering principle for the elements than atomic weight. This research is also important because it allowed others to determine just how many elements were still awaiting discovery between the limits of the naturally occurring elements (hydrogen and uranium).

Moseley received his training at the University of Manchester as a student of Rutherford. Moseley's experiments consisted of bouncing light off the surface of samples of various elements and recording the characteristic X-ray frequency that each one emitted. Such emissions occur because an inner electron is ejected from the atom, causing an outer electron to fill the empty space, in a process which is accompanied by the emission of X-rays.

Moseley first selected fourteen elements, nine of which, titanium to zinc, formed a continuous sequence of elements in the periodic table. What he discovered was that a plot of the emitted X-ray frequency against the square of an integer representing the position of each element in the periodic table produced a straight-line graph. Here was confirmation of van den Broek's hypothesis that the elements can be ordered by means of a sequence of integers, later called atomic number, one for each element, starting with H = 1, He = 2, and so on.

In a second paper, he extended this relationship to a further thirty elements, thus solidifying its status further. It then became a relatively simple matter for Moseley to verify whether the declarations of many newly claimed elements were valid or not. For example, the Japanese chemist Seiji Ogawa had announced that he had isolated an element to fill the space below manganese in the periodic table. Moseley measured the frequency of X-rays that Ogawa's sample produced when bombarded with electrons and found that it did not correspond to the value expected of element 43.

While chemists had been using atomic weights to order the elements there had been a great deal of uncertainty about just how many elements remained to be discovered. This was due to the irregular gaps that occurred between the values of the atomic weights of successive elements in the periodic table. This complication disappeared when the switch was made to using atomic number. Now the gaps between successive elements became perfectly regular, namely one unit of atomic number.

After Moseley died, other chemists and physicists used his method and found that the remaining unknown elements were those with atomic numbers of 43, 61, 72, 75, 85, 87, and 91 among the evenly spaced sequence of atomic numbers. It was not until 1945 when the last of these gaps was filled, following the synthesis of element number 61, or promethium.

Isotopes

The discovery of isotopes of any particular element is another key step in understanding the periodic table which occurred at the dawn of atomic physics. The term comes from *iso* (same) and *topos* (place) and is used to describe atomic species of any particular element which differ in weight and yet occupy the same place in the periodic table. The discovery came about partly as a matter of necessity. The new developments in atomic physics led to the discovery of a number of new elements such as Ra, Po, Rn, and Ac which easily assumed their rightful places in the periodic table. But in addition, thirty or so more apparently new elements were discovered over a short period of time. These new species were given provisional names like thorium emanation, radium emanation, actinium X, uranium X, thorium X, and so on, to indicate the elements which seemed to be producing them. The Xs denoted unknown species, which turned out to be isotopes of different elements in most cases. For example, uranium X was later recognized to be an isotope of thorium.

Some designers of periodic tables, like van den Broek, attempted to accommodate these new 'elements' into extended periodic tables as we saw earlier. Meanwhile, two Swedes, Daniel Strömholm and Theodor Svedberg, produced periodic tables in which some of these exotic new species were forced into the same place. For example, below the inert gas xenon, they placed radium emanation, actinium emanation, and thorium emanation. This seems to represent an anticipation of isotopy, but not a very clear recognition of the phenomenon.

In 1907, the year of Mendeleev's death, the American radiochemist Herbert McCoy concluded that 'radiothorium is entirely inseparable from thorium by chemical processes'. This was a key observation which was soon repeated in the case of many other pairs of substances that had originally been thought to be new elements. The full appreciation of such observations was made by Frederick Soddy, yet another former student of Rutherford.

To Soddy, the chemical inseparability meant only one thing, namely that these were two forms, or more, of the same chemical element. In 1913, he coined the term 'isotopes' to signify two or more atoms of the same element which were chemically completely inseparable, but which had different atomic weights. Chemical inseparability was also observed by Friedrich Paneth and Georg von Hevesy in the case of lead and 'radio lead', when

Rutherford had asked them to separate them chemically. After attempting this feat by twenty different chemical approaches, they were forced to admit complete failure. It was a failure in some respects but one that solidified further the notion of one element, lead in this case, occurring as chemically inseparable isotopes. But it was not all wasted work because Paneth and von Hevesy's efforts allowed them to establish a new technique for the radioactive labelling of molecules, which became the basis for a very useful subdiscipline with far-reaching applications, in areas like biochemistry and medical research.

In 1914, the case for isotopy gained even greater support from the work of T. W. Richards at Harvard, who set about measuring the atomic weights of two isotopes of the same element. He too selected lead, since this element was produced by a number of radioactive decay series. Not surprisingly, the lead atoms formed by these alternative pathways, involving quite different intermediate elements, resulted in the formation of atoms of lead differing by a large value of 0.75 of an atomic weight unit. This result was later extended by others to 0.85 of a unit.

Finally, the discovery of isotopes further clarified the occurrence of pair reversals, such as in the case of tellurium and iodine that had plagued Mendeleev. Tellurium has a higher atomic weight than iodine, even though it precedes it in the periodic table, because the weighted average of all the isotopes of tellurium happens to be a higher value than the weighted average of the isotopes of iodine. Atomic weight is thus seen as a contingent quantity depending upon the relative abundance of all the isotopes of an element. The more fundamental quantity, as far as the periodic table is concerned, is the van den Broek–Moseley atomic number or, as it was later realized, the number of protons in the nucleus. The identity of an element is captured by its atomic number and not its atomic weight, since the latter differs according to the particular sample from which the element has been isolated.

Whereas tellurium has a higher average atomic weight than iodine, its atomic number is one unit smaller. If one uses atomic number instead of atomic weight as an ordering principle for the elements, both tellurium and iodine fall into their appropriate groups in terms of chemical behaviour. It emerged, therefore, that pair reversals had been required only because an incorrect ordering principle had been used in all periodic tables prior to the turn of the 20th century.

Chapter 7
Electronic structure

Chapter 6 dealt mostly with discoveries in classical physics that did not require quantum theory. This was true of X-rays and radioactivity, which were largely studied without any quantum concepts, although the theory was later used to clarify certain aspects. In addition, the physics described in Chapter 6 was mostly about processes originating in the nucleus of the atom. Radioactivity is essentially about the break-up of the nucleus, and the transmutation of elements likewise takes place at the level of the nucleus. Moreover, atomic number is a property of the nuclei of atoms, and isotopes are distinguished by different masses of atoms of the same element, which are made up almost exclusively by the mass of their nuclei.

In this chapter, we venture into discoveries concerning the electrons in atoms, a study that did necessitate the use of quantum theory right from its early days. But first, we will need to say something of the origins of quantum theory itself. It all began before the turn of the 20th century in Germany, where a number of physicists were attempting to understand the behaviour of radiation held in a small cavity with blackened walls. The spectral behaviour of such 'black body radiation' was carefully recorded at different temperatures and attempts were then made to model the resulting graphs mathematically. The problem remained unsolved for a good deal of time until Max Planck, in the year 1900, made the bold assumption that the energy of this radiation consisted of discrete packets or 'quanta'. Planck himself seems to have been reluctant to accept the full significance of his new quantum theory and it was left to others to make some new applications of it.

The quantum theory asserted that energy comes in discrete bundles. This theory was successfully applied to the photoelectric effect in 1905 by none other than Albert Einstein, perhaps the most brilliant physicist of the 20th century. The outcome of his research was that light could be regarded as having a quantized, or particulate, nature. Nevertheless, Einstein soon began to regard quantum mechanics as an incomplete theory and was to maintain his criticism of it for the remainder of his life.

In 1913, the Dane Niels Bohr applied quantum theory to the hydrogen atom which he supposed, like Rutherford, to consist of a central nucleus with a circulating electron. Bohr assumed that the energy available to the electron occurred only in certain discrete values or, in pictorial terms, that the electron could exist in any number of shells or orbits around the nucleus. This model could explain, to some extent, a couple of features of the behaviour of the hydrogen atom and in fact atoms of any element. First, it explained why atoms of hydrogen which were exposed to a burst of electrical energy would result in a discontinuous spectrum in which only some rather specific frequencies were observed. Bohr reasoned that such behaviour came about when an electron underwent a transition from one allowed energy level to another one. Such a transition was accompanied by the release, or absorption, of the precise energy corresponding to the energy difference between the two energy levels in the atom.

Second, and less satisfactorily, the model explained why electrons did not lose energy and crash into the nucleus of any atom, as predicted from applying classical mechanics to a charge particle undergoing circular motion. Bohr's response was that the electrons would simply not lose energy provided that they remained in their fixed orbits. He also postulated that there was a lowest energy level beyond which the electron could not undergo any downward transitions.

Bohr then generalized his model to cover any many-electron atom rather than just hydrogen. First of all he assumed that a neutral atom with atomic number Z would also have Z electrons. He then set about trying to establish the way in which the electrons were arranged in any particular atom. Whereas the theoretical validity of making such a leap from one electron to many electrons was in question, this did not deter Bohr from forging ahead regardless. The electronic configurations that he arrived at are shown in Figure 23.

But Bohr's assignment of electrons to shells was not carried out on mathematical grounds, nor with any explicit help from quantum theory. Instead, Bohr appealed to chemical evidence such as the knowledge that atoms of the element boron can form three bonds, as do other elements in the boron group. An atom of boron therefore had to have three outer-shell electrons in order for this to be possible. But even with such a rudimentary and non-deductive theory, Bohr was providing the first successful electron-based explanation for why such elements as lithium, sodium, and potassium occur in the same group of the periodic table, and likewise for the membership of any group in the periodic table. In the case of lithium, sodium, and potassium, it is because each of these atoms has one electron that is in an outer shell that is set aside from the remaining electrons.

Bohr's theory had some other limitations, one of them being that it was strictly applicable only to one-electron atoms such as hydrogen or ions like He^{+}, Li^{2+}, Be^{3+}, etc. It was also found that some of the lines in the spectra for these 'hydrogenic' spectra broke up into unexpected pairs of lines. Arnold Sommerfeld working in Germany suggested that the nucleus might lie at one of the foci of an ellipse rather than at the heart of a circular atom.

1	H	1				
2	He	2				
3	Li	2	1			
4	Be	2	2			
5	B	2	3			
6	C	2	4			
7	N	4	3			
8	O	4	2	2		
9	F	4	4	1		
10	Ne	8	2			
11	Na	8	2	1		
12	Mg	8	2	2		
13	Al	8	2	3		
14	Si	8	2	4		
15	P	8	4	3		
16	S	8	4	2	2	
17	Cl	8	4	4	1	
18	Ar	8	8	2		
19	K	8	8	2	1	
20	Ca	8	8	2	2	
21	Sc	8	8	2	3	
22	Ti	8	8	2	4	
23	V	8	8	4	3	
24	Cr	8	8	2	2	2

23. Bohr's original 1913 scheme for electronic configurations of atoms.

His calculations showed that one had to introduce subshells within Bohr's main shells of electrons. Whereas Bohr's model was characterized by one quantum number denoting each of the separate shells or orbits, Sommerfeld's modified model required two quantum numbers to specify the elliptical path of the electron.

H	1				
He	2				
Li	2	1			
Be	2	2			
B	2	3			
C	2	4			
N	2	4	1		
O	2	4	2		
F	2	4	3		
Ne	2	4	4		
Na	2	4	4	1	
Mg	2	4	4	2	
Al	2	4	4	2	1
Si	2	4	4	4	
P	2	4	4	4	1
S	2	4	4	4	2
Cl	2	4	4	4	3
Ar	2	4	4	4	4

24. Bohr's 1923 electronic configurations based on two quantum numbers.

With these new quantum numbers Bohr was able to compile a more detailed set of electronic configurations in 1923, as shown in Figure 24.

A few years later the English physicist Edmund Stoner found that a third quantum number was also needed to specify some finer details of the spectrum of hydrogen and other atoms. Then in 1924 the Austrian-born theorist, Wolfgang Pauli, discovered the need for a fourth quantum number, which was identified with the concept of an electron adopting one of two values of a special kind of angular momentum. This new kind of motion was eventually called electron 'spin' even though electrons do not literally spin in the same way that the earth spins about an axis while also performing an orbital motion around the Sun.

The four quantum numbers are related to each other by a set of nested relationships. The range of values of the third quantum number depends on the value of the second quantum number which in turn depends on that of the first quantum number. Pauli's fourth quantum number is a little different since it can adopt values of +1/2 or –1/2 regardless of the values of the other three quantum numbers. The importance of the fourth quantum number, especially, is that its arrival provided a good explanation for why each electron shell can contain a certain number of electrons (2, 8, 18, 32, etc.), starting with the shell closest to the nucleus.

Here is how this scheme works. The first quantum number n can adopt any integral value starting with 1. The second quantum number, which is given the label l, can have any of the following values related to the values of n,

$$l = n-1, \ldots\ldots 0$$

In the case when n = 3, for example, l can take the values 2, 1, or 0. The third quantum number, labelled m_l, can adopt values related to those of the second quantum numbers as,

$$m = -l - (l-1), \ldots 0 \ldots (l-1), l$$

For example, if l = 2, the possible values of m_l are,

$$-2, -1, 0, +1, +2$$

Finally, the fourth quantum number labelled m_s can only take two possible values, either +1/2 or –1/2. There is therefore a hierarchy of related values for the four quantum numbers, which are used to describe any particular electron in an atom (Figure 25).

n	Possible values of l	Subshell designation	Possible values of m_l	Orbitals in subshell	Electrons in each shell
1	0	1s	0	1	2
2	0	2s	0	1	
	1	2p	1, 0, −1	3	8
3	0	3s	0	1	
	1	3p	1, 0, −1	3	
	2	3d	2, 1, 0, −1, −2	5	18
4	0	4s	0	1	
	1	4p	1, 0, −1	3	
	2	4d	2, 1, 0, −1, −2	5	
	3	4f	3, 2, 1, 0, −1, −2, −3	7	32

25. Combination of four quantum numbers to explain the total number of electrons in each shell.

As a result of this scheme it is clear why the third shell, for example, can contain a total of eighteen electrons. If the first quantum number, given by the shell number, is 3, there will be a total of $2 \times (3)^2$ or eighteen electrons in the third shell. The second quantum number l can take values of 2, 1, or 0. Each of these values of l will generate a number of possible values of m_l and each of these values will be multiplied by a factor of two since the fourth quantum number can adopt values of 1/2 or −1/2.

But the fact that the third shell can contain eighteen electrons does not strictly explain why it is that some of the periods in the periodic system contain eighteen places. It would be a rigorous explanation of this fact only if electron shells were filled in a strictly sequential manner. Although electron shells begin by filling in a sequential manner, this ceases to be the case starting with element number 19 or potassium. Configurations are built up starting with the 1s orbital which can contain two electrons, moving to the 2s orbital which likewise is filled with another two electrons. Then come the 2p orbitals which altogether contain a further six electrons, and so on. This process continues in a predictable manner up to element 18, argon, which has the configuration

$$1s^2, 2s^2, 2p^6, 3s^2, 3p^6.$$

One might expect that the configuration for the subsequent element, number 19, potassium, would be

$$1s^2, 2s^2, 2p^6, 3s^2, 3p^6, 3d^1,$$

where the final electron occupies the next subshell, which is labelled 3d. This would be expected because up to this point the pattern has been one of adding the differentiating electron to the next available orbital at increasing distances from the nucleus. However, experimental evidence shows that the configuration of potassium should be

$$1s^2, 2s^2, 2p^6, 3s^2, 3^6, 4s^1.$$

Similarly, in the case of element 20, calcium, the new electron also enters the 4s orbital. But in the next element, number 21, scandium, the configuration was generally thought to be (see Chapter 8)

$$1s^2, 2s^2, 2p^6, 3s^2, 3p^6, 4s^2, 3d^1.$$

This kind of skipping backwards and forwards among available orbitals as the electrons fill successive atoms recurs again several times. The identity of the differentiating electron as one moves through the periodic table is summarized in Figure 26.

As a consequence of this order of filling, successive periods in the periodic table contain the following number of elements, 2, 8, 8,

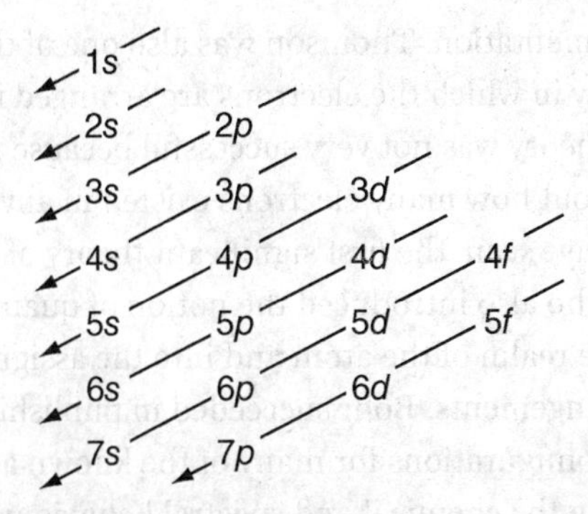

26. Approximate order of orbital filling. More strictly, order of differentiating electron. (see p. 109).

18, 18, 32, etc., thus showing a 'doubling' for each period except the first one.

Whereas the rule for the combination of four quantum numbers provides a rigorous explanation for the point at which shells close, it does not provide an equally rigorous explanation for the point at which periods close. Nevertheless, some rationalizations can be given for this order of filling, but such rationalizations tend to be somewhat dependent on the facts one is trying to explain. We know where the periods close because we know that the noble gases occur at elements 2, 10, 18, 36, 54, etc. Similarly, we have a knowledge of the order of orbital filling from observations but not from theory. The conclusion, seldom acknowledged in textbook accounts of the explanation of the periodic table, is that quantum physics does not fully explain the periodic table. Nobody has yet deduced the order of orbital filling from the principles of quantum mechanics. This is not to say that it might not be achieved in the future or that the order of electron filling is in any sense inherently irreducible to quantum physics.

Chemists and configurations

The discovery of the electron by J. J. Thomson in 1897 opened up all kinds of new explanations in physics as well as entirely new lines of experimentation. Thomson was also one of the first to discuss the way in which the electrons are arranged in atoms, although his theory was not very successful because not enough was known about how many electrons existed in any particular atom. As we have seen, the first significant theory of this kind was due to Bohr, who also introduced the notion of quantization of energy into the realm of the atom and into the assignment of electronic arrangements. Bohr succeeded in publishing a set of electronic configurations for many of the known atoms but only after consulting the chemical and spectral behaviour that had been gathered together by others over many years.

But what were the chemists doing at about this time? How did their attempts to exploit the electron compare with those of Bohr and other quantum physicists? To begin this survey requires going back in time to 1902, five years after the electron had been discovered. The American chemist G. N. Lewis, who was working in the Philippines at the time, sketched a diagram that is shown in Figure 27, the original copy of which still exists to this day. In it, he assumed that electrons occur at the corners of a cube and that as we move through the periodic table, one element at a time, a further electron is added to each corner. The choice of a cube may seem to be rather bizarre from a modern perspective since we now know that electrons move around the nucleus. But Lewis's model makes good sense in one important respect that is connected with the periodic table. Eight is the number of elements through which one must progress before a repetition in properties occurs.

Lewis was therefore suggesting that chemical periodicity and the properties of individual elements were governed by the number of electrons in the outermost electron cube around the nucleus of an atom. Although this is an incorrect model in that the electrons are regarded as being stationary, the choice of a cube is a natural and ingenious one which reflects the fact that chemical periodicity is based on intervals of eight elements.

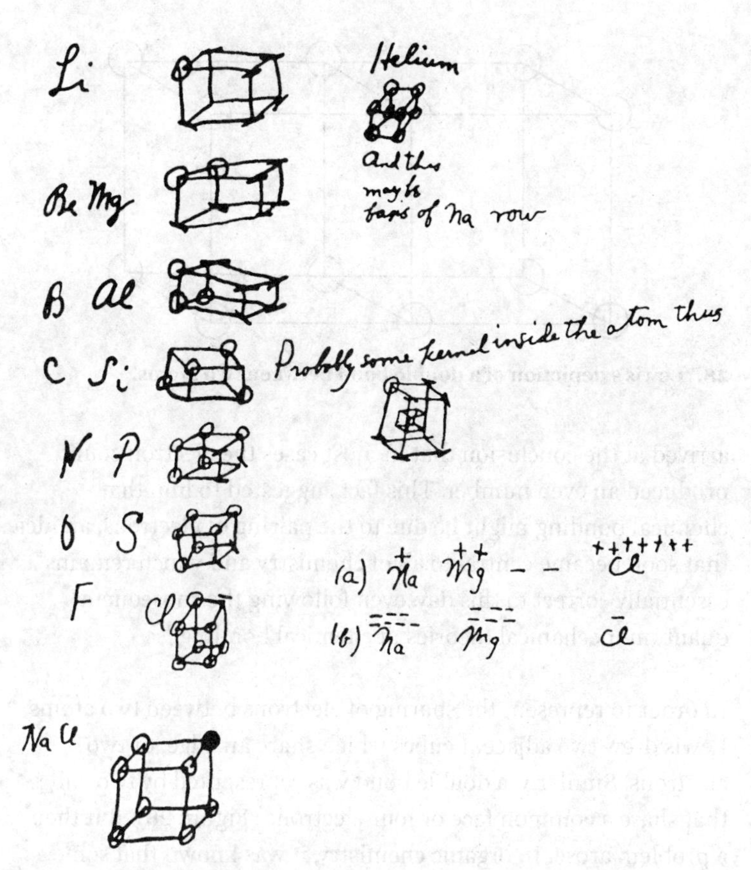

27. Lewis's sketch of atomic cubes.

In the same famous sketch, Lewis gave a diagram of how sodium and chlorine atoms can form a compound, after an electron is transferred from a sodium atom to one of chlorine, in order to occupy the place of the missing eighth electron in the outer cube of chlorine. Lewis then waited for a period of fourteen years before publishing these ideas as well as extending them to include another form of bonding, namely covalent bonding, that involves the sharing rather than the transfer of electrons between various atoms.

By considering many known compounds, and by counting the number of outer electrons that their atoms possessed, Lewis

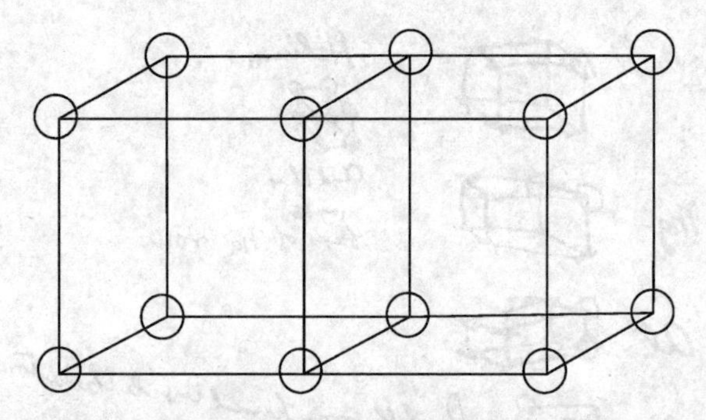

28. Lewis's depiction of a double bond between two atoms.

arrived at the conclusion that in most cases the electron count produced an even number. This fact suggested to him that chemical bonding might be due to the pairing of electrons, an idea that soon became central to all of chemistry and which remains essentially correct to this day, even following the subsequent quantum mechanical theories of chemical bonding.

In order to represent the sharing of electrons between two atoms, Lewis drew two adjacent cubes which share an edge, or two electrons. Similarly, a double bond was represented by two cubes that share a common face or four electrons (Figure 28). But then a problem arose. In organic chemistry, it was known that some compounds, such as acetylene, C_2H_2, contain a triple bond. Lewis realized that his model of electrons on the corners of a cube could not represent such triple bonds. In the same paper, he switched to a new model in which four pairs of electrons were located at the corners of a tetrahedron instead of a cube. A triple bond was then rendered by two tetrahedra that share a common face.

In the same paper, Lewis also returned to the question of electronic configurations of atoms and now extended his arrangements to include twenty-nine elements as shown in Figure 29. Before moving on to other contributors it is worth pausing to consider that G. N. Lewis is possibly the most significant chemist of the 20th century not to have been awarded a Nobel Prize. This partly

1	2	3	4	5	6	7
H						
Li	Be	B	C	N	O	F
Na	Mg	Al	Si	P	S	Cl
K	Ca	Sc		As	Se	Br
Rb	Sr			Sb	Te	I
CS	Ba			Bi		

29. Lewis's outer electronic structures for twenty-nine elements. The number at the head of the column represents the number of outer shell electrons for each atom (compiled by the author).

stems from his having died an early death in his own laboratory as a result of hydrogen cyanide poisoning and the fact that Nobel Prizes are not awarded posthumously. Last but not least, Lewis made too many enemies in the course of his academic career and so did not endear himself to colleagues who might otherwise have nominated him for the award.

Lewis's ideas were extended and popularized by the American industrial chemist Irving Langmuir. Whereas Lewis had assigned electronic configurations to only twenty-nine elements, Langmuir set out to complete the task. While Lewis had refrained from assigning configurations to the atoms of d-block metal elements, Langmuir gave the following list in a paper dated 1919:

Sc	Ti	V	Cr	Mn	Fe	Co	Ni	Cu	Zn
3	4	5	6	7	8	9	10	11	12

As Lewis had done before him, Langmuir used the chemical properties of the elements to guide him into these assignments and not any arguments from quantum theory (Figure 30). Not surprisingly, these chemists were able to improve upon the electronic configurations which physicists like Bohr were groping towards.

TABLE I.

Classification of the Elements According to the Arrangement of Their Electrons.

Layer.	N	E = 0	1	2	3	4	5	6	7	8	9	10
I			H	He								
IIa	2	He	Li	Be	B	C	N	O	F	Ne		
IIb	10	Ne	Na	Mg	Al	Si	P	S	Cl	A		
IIIa	18	A	K	Ca	Sc	Ti	V	Cr	Mn	Fe	Co	Ni
			11	12	13	14	15	16	17	18		
IIIa	28	Niβ	Cu	Zn	Ga	Ge	As	Se	Br	Kr		
IIIb	36	Kr	Rb	Sr	Y	Zr	Cb	Mo	43	Ru	Rh	Pd
			11	12	13	14	15	16	17	18		
IIIb	46	Pdβ	Ag	Cd	In	Sn	Sb	Te	I	Xe		
IVa	54	Xe	Cs	Ba	La	Ce	Pr	Nd	61	Sa	Eu	Gd
			11	12	13	14	15	16	17	18		
IVa			Tb	Ho	Dy	Er	Tm	Tm_2	Yb	Lu		
		14	15	16	17	18	19	20	21	22	23	24
IVa	68	Erβ	Tmβ	Tm_2β	Ybβ	Luβ	Ta	W	75	Os	Ir	Pt
			25	26	27	28	29	30	31	32		
IVa	78	Ptβ	Au	Hg	Tl	Pb	Bi	RaF	85	Nt		
IVb	86	Nt	87	Ra	Ac	Th	Ux_2	U				

30. Langmuir's periodic table.

In 1921, the English chemist Charles Bury, working at the University College of Wales in Aberystwyth, challenged an idea that was implicit in the work of both Lewis and Langmuir. This is the assumption that electron shells are filled sequentially as one moves through the periodic table, adding an extra electron for each new element encountered. Bury claimed that his own refined set of electronic configurations provided better agreement with the known chemical facts.

To conclude, physicists provided a great impetus to attempts to understand the underlying basis of the periodic table, but the chemists of the day were in many cases able to apply the new physical ideas, such as electronic configurations, to better effect.

Chapter 8
Quantum mechanics

Chapter 7 considered the influence of early quantum theory, especially that of Bohr (Figure 31), on the explanation of the periodic table. This explanation culminated in Pauli's contribution: his introduction of a fourth quantum number and the exclusion principle that now bears his name. It then became possible to explain why each shell around the atom can contain a particular number of electrons (first shell: 2, second shell: 8, third shell: 18, etc.). If one assumes the correct order of filling of orbitals within these shells one can now explain the fact that the period lengths are actually: 2, 8, 8, 18, 18, etc. But any worthy explanation of the periodic table should be capable of deriving this sequence of values from first principles, *without* assuming the observed order of orbital filling.

The quantum theory of Bohr, even when augmented by Pauli's contributions, was therefore only a stepping stone to a more advanced theory. The Bohr–Pauli version is usually called quantum theory, or sometimes the old quantum theory to distinguish it from the later development which was to take place in 1925 and 1926 that became known as quantum mechanics. The use of the word 'theory' to denote the older version is a little unfortunate because it reinforces the common misconception that a theory is somewhat vague and on its way to becoming a scientific law or some other more solid piece of knowledge.

31. Niels Bohr.

But in science, a theory is a highly supported, although never proven, body of knowledge which if anything possesses a more elevated status than scientific laws. Many different laws are often subsumed within one over-arching 'theory'. So quantum mechanics, the successor theory, was every bit as much a 'theory' as was Bohr's old quantum theory, even though it was a more general and more successful theory.

Bohr's old quantum theory had several shortcomings, including the fact that it could not explain chemical bonding, but this all changed following the advent of quantum mechanics. It then became possible to go beyond the picaresque idea of G. N. Lewis that bonding resulted from the mere sharing of electrons between two or more atoms in a molecule. According to quantum mechanics, electrons behave as much as waves as they do as particles. By writing a wave equation for the motion of electrons around the nucleus, the Austrian Erwin Schrödinger was able to take a decisive step forward. The solutions of Schrödinger's equation represent the possible quantum states that the electrons in an atom can find themselves in. A little later two physicists, Friedrich Hund and Robert Mulliken, independently developed the molecular orbital theory in which bonding was found to occur as a result of constructive and destructive interference between electron waves on each atom in a molecule.

But we need to return to atoms and the periodic table. According to Bohr's theory the only atoms whose energy levels could be calculated were those consisting of just one electron, which includes the H atom and one-electron ions such as He^{+1}, Li^{+2}, Be^{+3}, etc. In the case of many-electron atoms, where 'many' means more than one, Bohr's old quantum theory was impotent. Conversely, using the new quantum mechanics many-electron atoms could be handled by theoreticians, although admittedly in an approximate fashion instead of exactly. This is due to a mathematical limitation. Any system with more than one electron consists of a so-called 'many-body problem', and such problems admit only approximate solutions.

So the energies of the quantum states for any many-electron atom can be approximately calculated from first principles, although there is extremely good agreement with observed energy values. Nevertheless, some global aspects of the periodic table have still not been derived from first principles to this day. For example, the order of orbital filling that was mentioned earlier is a case in point.

Atomic orbitals which make up subshells and shells in any atom are filled in order of increasing values of the sum of the first two quantum numbers denoting any particular orbital. This fact is summarized by the $n + l$ or Madelung rule (after Erwin Madelung). Orbitals fill in such a way that the quantity $n + l$ increases gradually, starting with a value of 1 for the 1s orbital.

By following the diagonal arrows starting at the 1s orbital, and then by moving to the next diagonal set of lines, one obtains the order of orbital filling, which is as follows (Figure 26):

$$1s < 2s < 2p < 3s < 3p < 4s < 3d < 4p < 5s \text{ etc.}$$

In spite of its many dazzling successes, quantum mechanics has still not succeeded in deriving this sequence of $n + l$ values in a purely theoretical fashion. This is not to diminish the successes of the newer theory, which are truly impressive. It is just to point out that a feature remains to be elucidated. Perhaps a quantum physicist may succeed in deriving the Madelung rule in the future, or perhaps it may require an even more powerful future theory to do so. I am not trying to suggest some kind of inherent and 'spooky' irreducibility of chemical phenomena to quantum physics, but am just concentrating on what has actually been achieved so far in the context of the periodic table.

But let's turn to what quantum mechanics has illuminated regarding chemical periodicity. As was mentioned earlier, the Schrödinger equation can be written down and solved for any atom system in the periodic table without any experimental input whatsoever. For example, physicists and theoretical chemists have succeeded in calculating all the ionization energies of the atoms in the periodic table. When these calculations are compared with experimental values a remarkable degree of agreement is found (Figure 32).

Now it so happens that ionization energy is one of many properties of atoms which show marked periodicity. As we

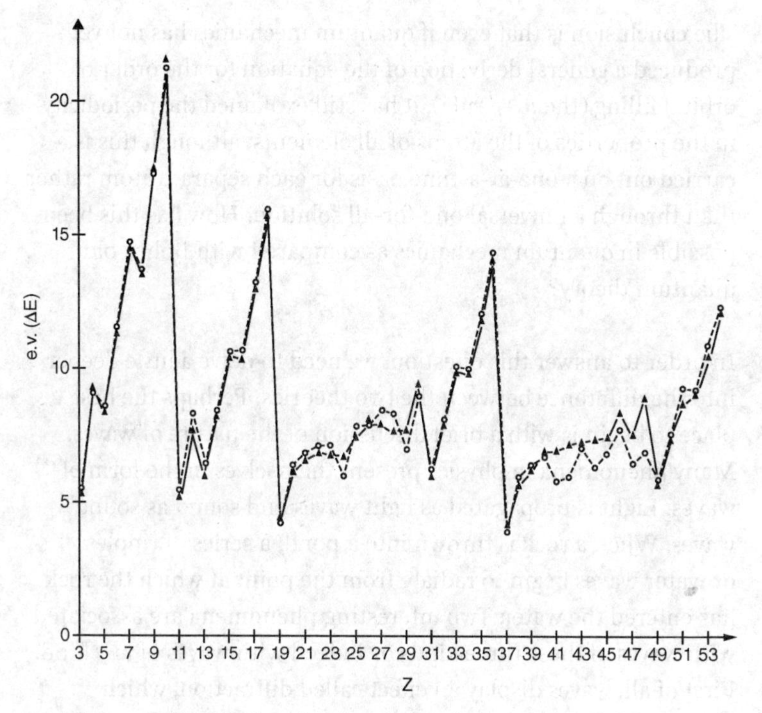

32. Theoretically calculated (triangles) and observed ionization energies (circles) of elements 1 to 53.

increase the atomic number starting with $Z = 1$ for hydrogen, the ionization energy shows an increase until we reach the very next element, helium. This is followed by a sharp decrease before we arrive at $Z = 3$, lithium. Subsequent values of ionization energy are then found to increase, broadly speaking, until we reach an element that is chemically analogous to helium, namely neon, which is another noble gas. This pattern of overall increase in ionization energy then recurs across all periods of the periodic table. And for each period the minimum value occurs for elements in group 1 such as lithium, sodium, and potassium. Maximum values, as we have just seen, occur at the noble gases such as helium, neon, argon, krypton, and xenon. Figure 32 also shows the curve obtained when theoretically computed values are connected together.

The conclusion is that even if quantum mechanics has not yet produced a general derivation of the equation for the order of orbital filling (the $n + l$ rule), it has still explained the periodicity in the properties of the atoms of all elements, although this is carried out on a one-at-a-time basis for each separate atom rather than through a universal one-for-all solution. How has this been possible in quantum mechanics as compared with Bohr's old quantum theory?

In order to answer this question, we need to delve a little deeper into the difference between the two theories. Perhaps the best place to begin is with a brief discussion of the nature of waves. Many phenomena in physics present themselves in the form of waves. Light is propagated as light waves and sound as sound waves. When a rock is thrown into a pond, a series of ripples or water waves begin to radiate from the point at which the rock has entered the water. Two interesting phenomena are associated with waves of these three kinds or indeed with waves of any kind. First of all, waves display an effect called diffraction, which describes the way in which they tend to spread when they pass through an aperture or around any obstacle.

In addition, two or more waves arriving at a screen together are said to be 'in phase' and produce an effect called constructive interference which leads to an overall increase in intensity over and above the intensities of the two separate waves. Conversely, two waves arriving at a screen in an out-of-phase fashion bring about destructive interference, the result of which is a cancellation of wave intensity. In the early 1920s, for reasons that we cannot go into here, it was suspected that particles like electrons might just behave as waves under certain conditions. Briefly put, this idea was a logical inversion of Einstein's work on the photoelectric effect, in which it had been established that light waves also behave as particles.

In order to test whether particles such as electrons really behave as waves, it became necessary to see whether electrons might also produce diffraction and interference effects like all other kinds of waves can. Surprising as it might seem, such experiments succeeded and the wave nature of electrons was established once and for all. In addition, a beam of electrons directed at a single crystal of metallic nickel was found to produce a series of concentric rings, that is to say a diffraction pattern was formed due to electron waves spreading around the crystal rather than just bouncing off it.

From this point onwards, electrons, and other fundamental particles, had to be regarded as having a kind of schizophrenic nature, behaving as both particles and waves. More specifically, the news that electrons behave as waves quickly reached the wider community of theoretical physicists, including Erwin Schrödinger. This Austrian-born theorist set about using the well-trodden mathematical techniques for modelling waves in order to describe the behaviour of the electron in a hydrogen atom. His assumptions were first that electrons behaved as waves, and second that the potential energy of the electron consisted of the energy with which it was attracted by the nucleus. Schrödinger solved his equation by doing what mathematical physicists always do in trying to solve differential equations of this kind: they apply boundary conditions.

Consider a guitar string that is bound at both ends, called the nut and the bridge respectively. Now strike the string to play an open string. As Figure 33 shows, the string will vibrate in an up and down motion in the form of half of one complete wave. But how else can the string vibrate? The answer is that it can also vibrate

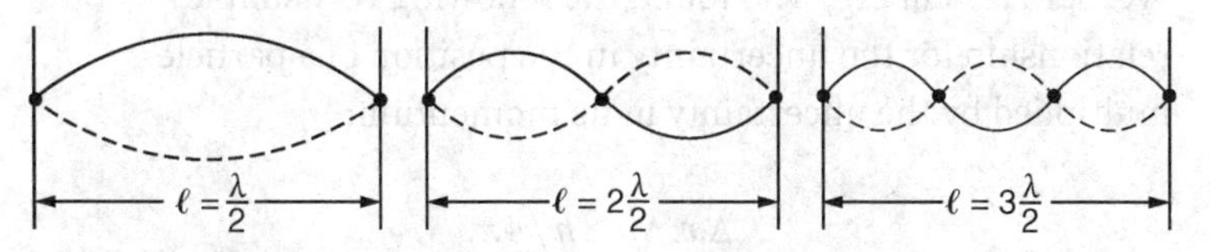

33. Boundary conditions and the emergence of quantization.

as two half-wavelengths (a whole wavelength). This can be achieved on the guitar string by lightly placing a finger on the string, halfway along the fret board (twelfth fret) and releasing it a fraction of a second after the string has been plucked in the region of the soundhole by the other hand. The sound produced, if this is done properly, is a rather pleasant sounding bell-like tone which musicians call a harmonic. Additional modes in which the string can vibrate consist of 3, 4, 5, etc. half waves. The string may only vibrate as a whole number of half waves. For example, there is no possible mode consisting of two and a half or three and one third half waves.

In mathematical terms, as well as physical terms, what has happened is that imposing boundary conditions (the two fixed points in the case of the guitar string) has resulted in a set of motions that are characterized by whole numbers. This amounts to nothing less than quantization, or the restriction to whole-number multiples of some value. By analogy, when Schrödinger applied such mathematical boundary conditions to his equation, the energies that he calculated for the electron turned out to be quantized, something that Bohr had been forced to smuggle into his equations from the outset. In Schrödinger's case, there is a substantial improvement in that he succeeded in deriving quantization of energy rather than merely introducing it artificially. This is the mark of a deeper treatment in which quantization emerges as a natural aspect of the theory.

Then there was another important ingredient that distinguished the new quantum mechanics from Bohr's old quantum theory. This second ingredient was discovered by the German physicist Werner Heisenberg, who found the following very simple relationship for the uncertainty in the position of a particle multiplied by the uncertainty in its momentum:

$$\Delta x . \Delta p \geq h / 4\pi .$$

This equation implies that we must abandon the common-sense notion that a particle like an electron has a simultaneously well-defined position and momentum. What Heisenberg's relation holds is that the more we pinpoint the position of the electron, the less we are able to pinpoint its momentum and vice versa. It is as though the motion of particles takes on a fuzzy or uncertain nature. Whereas Bohr's model had dealt with precise, planetary-like orbits of electrons circling the nucleus, the new view in quantum mechanics is that we may no longer speak of definite orbits for electrons. Instead, the theory retreats to talking in terms of probabilities (uncertainties) instead of certainties. When this view is combined with the idea that electrons are waves, the picture that emerges is radically different from Bohr's model.

In quantum mechanics, the electron is considered to be spread throughout a spherical shell. It is as though the familiar particle has been converted into a gas which spreads to fill the entire contents of a sphere which corresponds roughly to a three-dimensional version of Bohr's two-dimensional orbit. In addition, the quantum mechanical sphere of charge does not have a sharp edge to it because of Heisenberg uncertainty. The first, or 1s orbital, which we have been discussing, is in fact a volume that is associated with a 90 per cent probability of finding what we normally regard as a localized electron. And this is only the first of many solutions to the Schrödinger wave equation. Larger orbitals, as we move further from the nucleus, begin to have shapes other than spherical shells (Figure 34).

Now to consider the broader picture concerning quantum theory, quantum mechanics, and the periodic table. As we saw in Chapter 7, Bohr's original introduction of the concept of quantization into the study of atomic structure was carried out for the hydrogen atom. But in the same set of original papers, he began to explain the form of the periodic table by assuming that energy

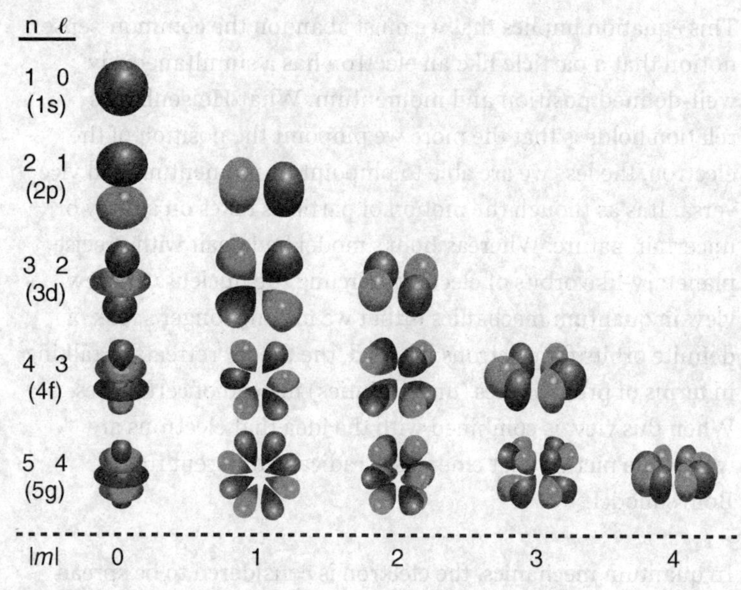

34. Orbitals s, p, d, f and g.

quantization also took place in many-electron atoms. The subsequent introduction of three further quantum numbers, in addition to Bohr's first quantum number, was also carried out in order to better explain the periodic table. And even before Bohr's venture into quantization concepts, the leading atomic physicist of the day, J. J. Thomson, had begun speculating about the arrangements of electrons in the various atoms that make up the periodic table.

The periodic table has therefore served as a testing ground for the theories of atomic physics and for many early aspects of quantum theory and the later quantum mechanics. The situation that exists today is that chemistry, and in particular the periodic table, is regarded as being fully explained by quantum mechanics. Even though this is not quite the case, the explanatory role that the theory continues to play is quite undeniable. But what seems to be forgotten in the current reductionist climate is that the periodic table led to the development of many aspects of modern quantum mechanics.

Postscript on the 4s–3d conundrum

For many years it was believed that a paradox exists regarding several pairs of atomic orbitals such as 4s and 3d. As we saw earlier, the 4s orbital is preferentially occupied with electrons in the atoms of potassium and calcium. It has generally been assumed that this would also be the case in the subsequent atoms starting with scandium, which has an atomic number of 21. This mistaken belief led to the following paradox which textbooks were unable to resolve in a satisfactory fashion. If the order of occupation in scandium is indeed

$$1s^2\ 2s^2\ 2p^6\ 3s^2\ 3p^6\ 4s^2\ 3d^1$$

as stated on page 92, this raises the question of why the ionization of this atom concerns the removal of a 4s electron rather than one from a 3d orbital. The issue has been resolved recently, since the publication of the first edition of this book.

In fact the correct order of orbital occupation in scandium is:

$$1s^2\ 2s^2\ 2p^6\ 3s^2\ 3p^6\ 3d^1\ 4s^2.$$

Consequently, the preferential ionization of a 4s electron ceases to raise any problems. I have dubbed the former incorrect thinking the 'sloppy Aufbau', in which it was thought that the configuration of scandium and subsequent transition metal atoms were built up by the mere addition of a further electron to the configuration of the previous atom of calcium in which the 4s orbital is indeed preferentially occupied. It is now recognized that this is not the case and indeed the cause of the previous conundrum that was believed to exist (see Further reading for details).

Anomalous configurations and a second recent finding

Another previously mysterious topic having to do with atomic configurations and the periodic table has also recently been resolved. This concerns the irregular or anomalous manner in which as many as twenty transition metal atoms seem to behave as we move through the periodic table. In the first transition series the atoms of chromium and copper have just one electron in their outermost s-orbital, by contrast with the remaining eight elements of this series that have two such electrons. In the second series, as many as six out of ten elements show similar anomalies.

The fully satisfactory explanation for these facts has now been provided by the theorist Eugen Schwarz. He has shown that if suitable weighted averages over the appropriate spectral energy levels are taken, the commonly discussed anomalies can now be regarded as reflecting the relative stabilities of the s^1 and s^2 configurations of these atoms. Looked at in this way there are in fact no anomalous configurations but just various configurations all competing to act as the lowest energy configuration (see Further reading for details).

Chapter 9
Modern alchemy: from missing elements to synthetic elements

The periodic table consists of about ninety elements which occur naturally, ending with element 92, uranium. I say about ninety because one or two elements like technetium were first created artificially and only later found to occur naturally on earth.

Chemists and physicists have succeeded in synthesizing some of the elements that were missing between hydrogen (1) and uranium (92). In addition they have synthesized a further twenty-five or so new elements beyond uranium, although, again, one or two of these (like neptunium and plutonium) were later found to exist naturally in exceedingly small amounts.

All elements up to and including element 118 have now been synthesized. The most recent additions have included elements 113, 115, and 117 as well as 118 that were officially named nihonium, moscovium, tennessine, and oganesson respectively.

The synthesis of many elements involved starting with a particular nucleus and subjecting it to bombardment with small particles with the aim of increasing the atomic number and hence changing the identity of the nucleus in question. More recently, the method of synthesis has changed to involve the collision of nuclei of considerable weights but still with the aim of forming a larger nucleus with a larger *Z* value.

There is a fundamental sense in which all these syntheses are descended from a crucial experiment, conducted in 1919 by Rutherford and Soddy at the University of Manchester. What Rutherford and colleagues did was to bombard nuclei of nitrogen with alpha particles (helium nuclei), with the result that the nitrogen nucleus was transformed into that of another element. Although they did not realize it initially, the reaction had produced an isotope of oxygen. Rutherford had achieved the first ever transmutation of one element into a completely different one. The dream of the ancient alchemists had become a reality, and this general process has continued to yield new elements right up to the present time.

However, this reaction did not produce a new element but just an unusual isotope of an existing element. Rutherford used alpha particles produced by the radioactive decay of other unstable nuclei such as uranium. It soon emerged that similar transmutations could be carried out with target atoms other than nitrogen but extending only as far as calcium, which has the atomic number 20. If heavier nuclei were to be transmuted, it would require more energetic projectiles than naturally produced alpha particles because the incoming projectile had to overcome the repulsion of the highly charged target nucleus.

The situation changed in the 1930s following the invention of the cyclotron by Ernest Lawrence at the University of California, Berkeley. This machine made it possible to accelerate alpha particles to hundreds and even thousands of times the speed possessed by naturally produced alpha particles. In addition, another projectile particle, the neutron, was discovered in 1932, having the added advantage of possessing a zero electric charge, which meant that it could penetrate a target atom without suffering any repulsion from charged protons inside the nucleus.

Missing elements

In the mid-1930s, four gaps remained to be filled in the then existing periodic table. They consisted of the elements with atomic numbers 43, 61, 85, and 87. Interestingly, the existence of three of these elements had been clearly predicted by Mendeleev many years before and called by him eka-manganese (43), eka-iodine (85), and eka-caesium (87). Three of these four missing elements were first discovered as a result of their being artificially synthesized in the 20th century.

In 1937, experiments were conducted at the Berkeley cyclotron facility in which a molybdenum target was bombarded with deuterium projectiles (isotopes of hydrogen with twice the mass of the more abundant isotope). One of the researchers was Emilio Segrè, a postdoctoral fellow from Sicily who then took the irradiated plates back to his native country. In Palermo, Segrè and Perrier analysed the plates and were able to confirm the formation of a new element, number 43, which they subsequently named technetium.

This was the first completely new element obtained by transmutation, eighteen years after Rutherford's classic experiment had demonstrated the possibility. Traces of the new element, technetium, were later found to occur naturally in the earth in vanishingly small quantities.

Element 85, or Mendeleev's eka-iodine, was the second missing element discovered as a result of artificial synthesis. Given its atomic number of 85, it could have been formed either from polonium (84) or bismuth (83). Polonium is very unstable and radioactive, and so attention turned to bismuth. And given that the atomic number of bismuth is two short of that of element 85, the obvious bombarding projectile would once again be an alpha particle. In 1940, Dale Corson, Mackenzie, and Segrè, who had by then permanently settled in the USA, conducted just such an experiment in Berkeley and obtained an isotope of element 85 with a half-life of 8.3 hours. They named it astatine from the Greek *astatos*, meaning 'unstable'. The third of the missing elements to be artificially synthesized was element 61, also at the Berkeley cyclotron, this time by the team consisting of Jacob Marinsky, Lawrence Glendenin, and Charles Coryell. The reaction involved the collision of neodymium atoms with deuterium atoms.

Just to complete the story, the fourth missing element had already been discovered by the French chemist Marguerite Perey, in 1939. She had started working as a laboratory assistant to Madame Curie and did not even have an undergraduate degree at the time of her discovery. She named her newly discovered element francium after her native country. This element did not require artificial synthesis but was found to be the by-product of the natural radioactive disintegration of actinium. Perey eventually rose to full professor and the directorship of the main institution for nuclear chemistry in France.

Transuranium elements

We finally come to the synthesis of the elements beyond the original 1 to 92. In 1934, three years before the synthesis of technetium, Enrico Fermi, working in Rome, began bombarding element targets with neutrons in the hope of synthesizing transuranium elements. Fermi believed that he had succeeded in producing two such elements, which he promptly named ausonium (93) and hesperium (94). But it was not to be.

After announcing the findings in his Nobel Prize acceptance speech of the same year, he quickly retracted the claim when it came to producing a written version of his presentation. The explanation for the erroneous claims emerged one year later, in 1938, when Otto Hahn, Fritz Strassmann, and Lise Meitner discovered nuclear fission. It became clear that, on collision with a neutron, the uranium nucleus, for example, could break up to form two middle-sized nuclei rather than a larger one. For example, uranium was capable of forming barium and krypton.

Fermi and his collaborators had been observing such products of nuclear fission processes instead of forming heavier nuclei as they first believed.

Real transuranium elements

The true identification of element 93 was finally carried out, in 1939, by Edwin McMillan and collaborators at Berkeley. It was named neptunium because it followed uranium in the periodic table just as the planet Neptune follows Uranus in terms of its distance from the Sun. A chemist in the team, Philip Abelson, discovered that element 93 did not behave as eka-rhenium was expected to, based on its presumed position in the periodic table. It was on the basis of this, and similar findings on element 94, or plutonium, that Glenn Seaborg was to propose a major modification to the periodic table (see Chapter 1). As a result, the elements from actinium (89) onwards were no longer regarded as transition metals but as analogues of the lanthanoid series. Consequently, there was no need for elements like 93 and 94 to behave like eka-rhenium and eka-osmium, since they had been moved to different places on the revised periodic table.

The synthesis of elements 94 to 97 inclusive, named plutonium, americium, curium, and berkelium, took place in the remaining years of the 1940s, and number 98, or californium, arrived in 1950. But this sequence looked as though it was about to end, since the heavier the nucleus, the more unstable it is, generally speaking. The problem became one of needing to accumulate enough target material in the hope of bombarding it with neutrons to transform the element into a heavier one. At this point, serendipity intervened. In 1952, a thermonuclear test explosion, code name Mike, was carried out close to the Marshall Islands in the Pacific Ocean. One of the outcomes was that intense streams of neutrons were produced, thus enabling such reactions as would not have been possible otherwise at this time. For example, fifteen neutrons can collide with the U-238 isotope to form U-253, which subsequently undergoes the loss of seven beta particles, resulting in the formation of element 99, named einsteinium.

$$^{238}_{92}U + 15\,^{1}_{0}n \rightarrow \,^{253}_{92}U \rightarrow \,^{253}_{99}Es + 7\,^{0}_{-1}\beta$$

Element 100, named fermium, was produced in a similar manner as a result of the high-neutron flux produced by the same explosion and as revealed by analysis of the soil from the nearby Pacific islands.

From 101 to 106

Advancing further along the sequence of ever heavier nuclei required a quite different approach given that beta decay does not take place for elements above $Z = 100$. Many technological innovations were developed, including the use of linear accelerators rather than cyclotrons, the former allowing researchers to accelerate highly intense beams of ions at well-defined energies. The projectile particles could also now be heavier than neutrons or alpha particles. During the Cold War, the only countries that possessed such facilities were the two superpowers, the USA and the Soviet Union.

In 1955, mendelevium, element 101, was produced in this way at the linear accelerator at Berkeley:

$$^{4}_{2}He + \,^{253}_{99}Es \rightarrow \,^{256}_{101}Md + \,^{1}_{0}n$$

The possible combinations of nuclei became more numerous. For example, element 104, or rutherfordium, was made in Berkeley through the following reaction:

$$ {}^{12}_{6}C + {}^{249}_{98}Cf \rightarrow {}^{257}_{104}Rf + 4\,{}^{1}_{0}n $$

In Dubna, Russia, a different isotope of the same element was created in the reaction:

$$ {}^{22}_{10}Ne + {}^{242}_{94}Pu \rightarrow {}^{259}_{104}Rf + 5\,{}^{1}_{0}n $$

In all, six elements, 101 to 106, were synthesized by this approach. Partly as a result of Cold War tension between the USA and the Soviet Union, claims for the synthesis of most of these elements were hotly disputed, and these disputes continued for many years. But having reached element 106, a new problem arose that necessitated yet another new approach. It was at this point that German scientists entered the field, with the establishment of the GSI (*Gesellschaft für Schwerionenforschung*, or the Society for Heavy Ion Research) in Darmstadt. The new technology was named 'cold fusion' but had nothing to do with the kind of cold fusion in a test tube as erroneously announced by the chemists Martin Fleischmann and Stanley Pons in 1989.

Cold fusion in the transuranium field is a technique whereby nuclei are made to collide with one another at lower speeds than were previously used. As a result, less energy is generated, and consequently there is a decreased possibility that the combined nucleus will fall apart. This technique was originally devised by Yuri Oganessian, a Soviet physicist, but was developed more fully in Germany.

Elements 107 onwards

In the early 1980s, elements 107 (bohrium), 108 (hassium), and 109 (meitnerium) were successfully synthesized in Germany using the cold fusion method. But then another roadblock became apparent. Many new ideas and techniques were tried, while collaboration between the USA, Germany, and what by now had become Russia began to develop after the fall of the Berlin Wall and the Soviet Union. In 1994, after ten years of stagnation, the GSI announced the synthesis of element 110 formed by the collision of lead and nickel ions.

The half-life of the resulting isotope was a mere 170 microseconds. The Germans called their element darmstadtium, not surprisingly given the previous naming of berkelium and dubnium by American and Russian teams respectively. A month later, the Germans had obtained element 111, which later became known as roentgenium after Röntgen, the discoverer of X-rays. February 1996 saw the synthesis of the next element in the sequence, 112, which was officially named copernicium in 2010.

Elements 113 to 118

Since 1997, several claims have been published for the synthesis of elements 113 all the way to 118, the latest being element 117, synthesized in 2010. This situation is well understood given that nuclei with an odd number of protons are invariably more unstable than those with an even number of protons. This difference in stability occurs because protons, like electrons, have a spin of one half and enter into energy orbitals, two by two, with opposite spins. It follows that even numbers of protons frequently produce total spins of zero and hence more stable nuclei than those with unpaired proton spins as occurs in nuclei with odd numbers of protons such as 115 or 117.

The synthesis of element 114 was much anticipated because it had been predicted for some time that it would represent the start of an 'island of stability', that is to say a portion of the table of nuclei with enhanced stability. Element 114 was first claimed by the Dubna lab in late 1998, but only definitely produced in further experiments in 1999 involving the collision of a plutonium target with calcium-48 ions. The labs at Berkeley and Darmstadt have recently confirmed this finding. At the time of writing, something like eighty decays involving element 114 have been reported, thirty of which come from the decay of heavier nuclei such as 116 and 118. The longest-lived isotope of element 114 has a mass of 289 and a half-life of about 2.6 seconds, in agreement with predictions that this element would show enhanced stability.

On 30 December 1998, the Dubna–Livermore labs published a joint paper, claiming element 118 had been observed as a result of the following reaction:

$$^{86}_{36}Kr + ^{208}_{82}Pb \rightarrow ^{293}_{118}Og + ^{1}_{0}n$$

After several failed attempts to reproduce this result in Japan, France, and Germany, the claim was officially retracted in July 2001. Much controversy followed, including the dismissal of one senior member of the research team who had published the original claim.

A couple of years later, new claims were announced from Dubna and were followed in 2006 by further claims by the Lawrence Livermore Laboratory in California. Collectively, the US and Russian scientists made a stronger claim, that they had detected four more decays of element 118 from the following reaction:

$$^{48}_{20}Ca + ^{249}_{98}Cf \rightarrow ^{294}_{118}Og + 3^{1}_{0}n$$

The researchers are highly confident that the results are reliable, since the chance that the detections were random events was estimated to be less than one in 100,000. Needless to say, no chemical experiments have yet been conducted on this element in view of the paucity of atoms produced and their very short lifetimes of less than one millisecond.

In 2010, an even more unstable element, number 117, tennessine, was synthesized and characterized by a large team of researchers working in Dubna, as well as several labs in the USA. The periodic table has reached an interesting point at which all 118 elements either exist in nature or have been created artificially in special experiments. This includes a remarkable twenty-six elements beyond the element uranium, which has been traditionally regarded as the last of the naturally occurring elements. The last four elements that served to complete the seventh period have been named nihonium (113), moscovium (115), tennessine (117), and oganesson (118). At the time of writing, there are even plans to attempt the creation of yet heavier elements such as 119 and 120, and there are no reasons for thinking that there should be any immediate end to the sequence of elements that can be formed.

Chemistry of the synthetic elements

The existence of superheavy elements raises an interesting new question and also a challenge to the periodic table. It also affords an intriguing new meeting point for theoretical predictions to be pitted against experimental findings. Theoretical calculations suggest that the effects of relativity become increasingly important as the nuclear charge of atoms increases. For example, the characteristic colour of gold, with a rather modest atomic number of 79, is now explained by appeal to relativity theory. The larger the nuclear charge, the faster the motion of inner-shell electrons. As a consequence of gaining relativistic speeds, such inner electrons are drawn closer to the nucleus, and this in turn has the effect of causing greater screening on the outermost electrons which determine the chemical properties of any particular element. It has been predicted that some atoms should behave chemically in a manner that is unexpected from their presumed positions in the periodic table.

Relativistic effects thus pose the latest challenge to test the universality of the periodic table. Such theoretical predictions have been published by various researchers over a period of many years, but it was only when elements 104 and 105, rutherfordium and dubnium respectively, were chemically examined that the situation reached something of a climax. It was found that the chemical behaviour of rutherfordium and dubnium was in fact rather different from what one would expect intuitively from where these elements lie in the periodic table. Rutherfordium and dubnium did not seem to behave like hafnium and tantalum, respectively, as they should have done.

For example, in 1990, K. R. Czerwinski, working at Berkeley, reported that the solution chemistry of element 104, or rutherfordium, differed from that of zirconium and hafnium, the two elements lying above it. Meanwhile, he also reported that rutherfordium's chemistry resembled that of the element plutonium which lies quite far away in the periodic table. As to dubnium, early studies showed that it too was not behaving like the element above it, namely tantalum (Figure 35). Instead, it showed greater similarities with the actinoid protactinium. In other experiments, rutherfordium and dubnium seemed to be behaving more like the two elements above hafnium and tantalum, namely zirconium and niobium.

It was only when the chemistry of the following elements, seaborgium (106) and bohrium (107), was examined that they showed that the expected periodic behaviour was resumed. The titles of the articles that announced these discoveries spoke for

3	4	5	6	7	8	9	10	11	12
Sc	Ti	V	Cr	Mn	Fe	Co	Ni	Cu	Zn
Y	Zr	Nb	Mo	Tc	Ru	Rh	Pd	Ag	Cd
Lu	Hf	Ta	W	Re	Os	Ir	Pt	Au	Hg
Lr	Rf	Db	Sg	Bh	Hs	Mt	Ds	Rg	Cn

35. Fragment of periodic table showing groups 3 to 12 inclusive.

themselves. They included, 'Oddly Ordinary Seaborgium' and 'Boring Bohrium', both references to the fact that it was business as usual for the periodic table. Even though relativistic effects should be even more pronounced for these two elements, the expected chemical behaviour seems to outweigh any such tendencies.

The fact that bohrium behaves as a good member of group 7 can be seen from the following argument that I have proposed. This approach also represents a kind of 'full circle' since it involves a triad of elements. As the reader may recall from Chapter 3, the discovery of triads was the very first hint of a numerical regularity relating the properties of elements within a common group. Figure 36 shows the data for measurements carried out on the standard sublimation enthalpies (energy required to convert a solid directly into a gas) of analogous compounds of technetium, rhenium, and bohrium with oxygen and chlorine.

Predicting the value for BhO_3Cl using the triad method gives 83 kJ/mol, or an error of only 6.7 per cent compared with the above experimental value of 89 kJ/mol. This fact lends further support to the notion that bohrium acts as a genuine group 7 element (Figure 35). The challenge to the periodic law from relativistic effects became even more pronounced in the case of number 112, copernicium. Once again, relativistic calculations indicated modified chemical behaviour to the extent that the element was thought to behave like a noble gas rather than like mercury below which it is placed in the periodic table. Experiments

TcO_3Cl = 49 kJ/mol
ReO_3Cl = 66 kJ/mol
BhO_3Cl = 89 kJ/mol

36. Sublimation enthalpies for elements in group 7 showing that element 107 is a genuine member of this group.

carried out on sublimation enthalpies on element 112 then showed that, contrary to earlier expectations, the element truly belongs in group 12 along with zinc, cadmium, and mercury.

Element 114 presented a similar story, with early calculations and experiments suggesting noble gas behaviour but more recent experiments supporting the notion that the element behaves like the metal lead, as expected from its position in group 14. The conclusion would seem to be that chemical periodicity is a remarkably robust phenomenon. Not even the powerful relativistic effects due to fast-moving electrons seem to be capable of toppling a simple scientific discovery that was made around 150 years ago, or at least the experiments conducted up to now are far from being conclusive on this question.

Chapter 10
Forms of the periodic table

Much has been said about the periodic table in previous chapters, but one important aspect has not yet been addressed. The question is why have so many periodic tables been published and why are there so many currently on offer in textbooks, articles, and on the Internet? Is there such a thing as an 'optimal' periodic table? Does such a question even make sense, and if so, what progress has been made towards identifying such an optimal table?

In his classic book on the history of the periodic table, Edward Mazurs included illustrations as well as references to about 700 periodic tables that have been published since the periodic table was first formulated in the 1860s. In the forty-five or so years that have elapsed since the publication of Mazurs' book, at least another 300 tables have appeared, especially if we include new periodic systems posted on the Internet. The very fact that so many periodic tables exist is something that requires an explanation. Of course, many of these tables may not have anything new to offer, and some are even inconsistent from a scientific point of view. But even if we were to eliminate these misleading tables, a very large number of tables still remain.

In Chapter 1, we saw that there are three basic forms of the periodic table, the short form, the medium-long form, and the long form. All three of these convey much the same information, although, as was noted, the grouping of elements with the same valence is treated differently in each of these formats. In addition, there are periodic tables that don't look like tables in the literal sense of having a rectangular shape. One important variation of this kind includes the circular and elliptical periodic systems which serve to highlight the continuity of the elements in a way that is perhaps better than the rectangular forms. In circular or elliptical systems, there are no discontinuities at the end of the periods as there are in the rectangular formats between, say, neon and sodium or argon and potassium. But the lengths of periods vary, unlike the periods on a clock face. As a result, the circular periodic table designer needs to accommodate the longer periods which have transition elements in some way. For example, Benfey's table (Figure 37) does this by using bulges for the transition metals which jut off the main circular system. Then there are three-dimensional periodic tables such as the one designed by Fernando Dufour from Montreal in Canada (Figure 38).

But I am going to suggest that all these variations only involve changing the portrayal of the periodic system and that there is no fundamental difference between them. What does constitute a major variation is the placement of one or more elements in a different group than where it is usually located in the conventional table. Before taking up this point, let me pause to discuss the design of the periodic table in general.

The periodic table is a deceptively simple concept, or at least that's how it appears. This is what invites amateur scientists to try their hand at developing new versions, which they often claim have properties that are superior to all previously published systems. Moreover, there have been a number of cases in which amateurs, or outsiders to the fields of chemistry and physics, have indeed made major contributions. For example, Anton van den Broek, who was mentioned in Chapter 6, was an economist who first

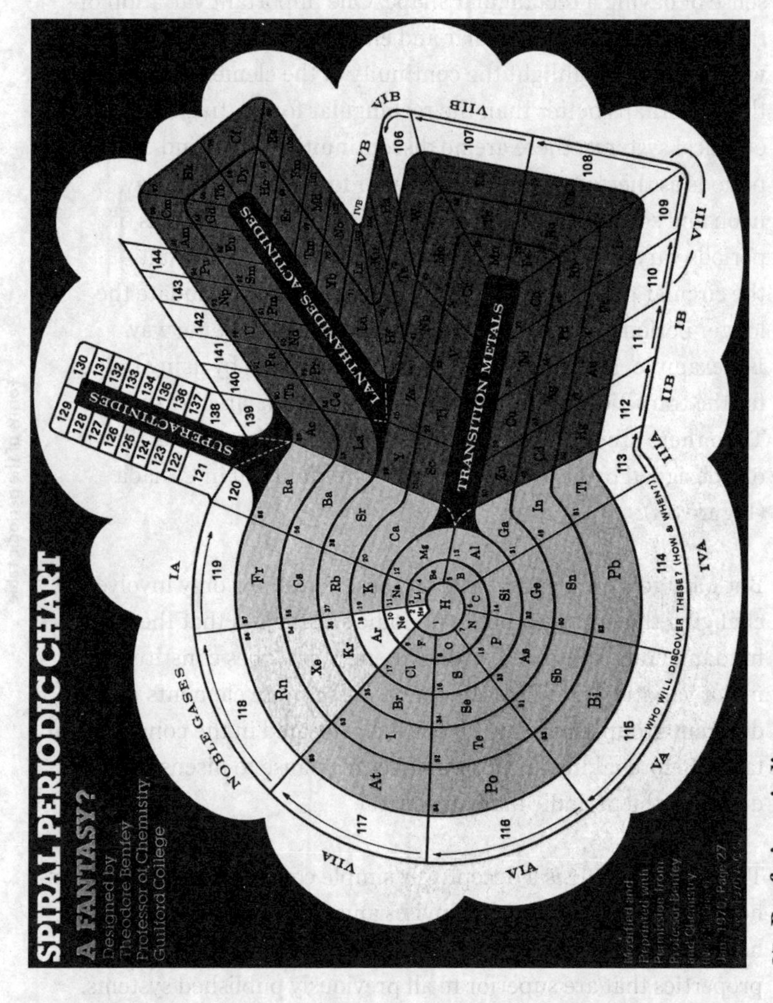

37. Benfey's periodic system.

38. Dufour's periodic tree.

came up with the notion of atomic number, an idea that he developed in several articles in such journals as *Nature*. A second example was Charles Janet, a French engineer, who in 1929 published the first known version of the left-step periodic system that continues to command much attention among periodic table amateurs as well as experts (Figure 39).

																														H	He
																														Li	Be
																								B	C	N	O	F	Ne	Na	Mg
																								Al	Si	P	S	Cl	Ar	K	Ca
														Sc	Ti	V	Cr	Mn	Fe	Co	Ni	Cu	Zn	Ga	Ge	As	Se	Br	Kr	Rb	Sr
														Y	Zr	Nb	Mo	Tc	Ru	Rh	Pd	Ag	Cd	In	Sn	Sb	Te	I	Xe	Cs	Ba
La	Ce	Pr	Nd	Pm	Sm	Eu	Gd	Tb	Dy	Ho	Er	Tm	Yb	Lu	Hf	Ta	W	Re	Os	Ir	Pt	Au	Hg	Tl	Pb	Bi	Po	At	Rn	Fr	Ra
Ac	Th	Pa	U	Np	Pu	Am	Cm	Bk	Cf	Es	Fm	Md	No	Lr	Rf	Db	Sg	Bh	Hs	Mt	Ds	Rg	Cn	Nh	Fl	Mc	Lv	Ts	Og	119	120

39. The left-step periodic table by Charles Janet.

Now, what about the other question that was raised earlier? Does it even make sense to be seeking an optimal periodic table, or are the amateurs deluding themselves along with the experts who devote time to this question? The answer to this question lies, I believe, in one's philosophical attitude towards the periodic system. If one believes that the approximate repetition in the properties of the elements is an objective fact about the natural world, then one is adopting the attitude of a realist. To such a person, the question of seeking an optimal periodic table does make perfect sense. The optimal periodic table is the one which best represents the facts of the matter concerning chemical periodicity, even if this optimal table may not yet have been formulated.

On the other hand, an instrumentalist or an anti-realist about the periodic table might believe that the periodicity of the elements is a property that is imposed on nature by human agents. If this is the case, there is no longer a burning desire to discover *the* optimal table since no such thing would even exist. For such a conventionalist or anti-realist, it does not matter precisely how the elements are represented since he or she believes that one is dealing with an artificial rather than a natural relationship among the elements.

Just to declare my own position, I am very much a realist when it comes to the periodic table. For example, it surprises me that many chemists take an anti-realistic stance concerning the periodic table; when asked whether the element hydrogen belongs to group 1 (alkali metals) or group 17 (the halogens), some chemists respond that it does not matter.

There are some final general issues to mention before we examine the details of alternative tables and possible optimal tables. One is the question of utility of this or that periodic table. Many scientists tend to favour one or other particular form of the periodic table because it may be of greater use to them in their scientific work as an astronomer, a geologist, a physicist, and so on.

These are the tables that are primarily based on utility. There are other tables that seek to highlight the 'truth' about the elements, for want of a better expression, rather than usefulness to some kinds of scientists. Needless to say, any quest for an optimal periodic table should avoid the question of utility, especially if it is utility to one particular scientific discipline or subdiscipline. Moreover, a table that strives for the truth about the elements, above all, will hopefully turn out to be useful to various disciplines if it succeeds in somehow capturing the true nature and relationship between the elements. But such usefulness would come as a bonus as it is not something that should determine the manner in which the optimal table is arrived at.

There is also the important question of symmetry. This too is a rather tricky issue. Many proponents of alternative periodic tables claim that their table is superior because the elements are represented in a more symmetrical, regular, or somehow more elegant or beautiful fashion. The question of symmetry and beauty in science has been much debated, but as with all aesthetic questions, what may appear beautiful to one person may not be regarded in the same way by somebody else. Also, one must beware of imposing beauty or regularity on nature where it might not actually be present. Too many proponents of alternative tables seem to argue exclusively about the regularity of their representation and sometimes forget that they are talking about the representation and not the chemical world itself.

Some specific cases

Having mentioned all these preliminaries, we can plunge into some proposed new tables, assuming of course that, as I am suggesting, it does make sense to seek an optimal periodic table. Let us begin with the left-step periodic table, which is one of those substantially altered periodic systems in which elements are placed in different groups than in the more conventional tables. The left-step table was first proposed by Charles Janet, in 1929, shortly after the development of quantum mechanics. However, it seems that Janet's proposal owes nothing to quantum mechanics but was based entirely on aesthetic grounds. It soon became clear that there are some major features of the left-step table which may correlate rather better with the quantum mechanical account of atoms than the conventional tables do.

The left-step table is obtained by moving the element helium from the top of the noble gases (group 18) to the top of the alkaline earth metals (group 2). The entire two groups on the left of this table are disconnected and moved to the right-hand edge to form a new table. In addition, the block of twenty-eight elements, the lanthanoids and actinoids, that are usually presented as a form of footnote to the periodic table, are moved to the left side of the new table. As a result of this move, these elements become fully incorporated into the periodic table to the left of the transition metal block.

Among the virtues of the new table is the fact that the overall shape is more regular and more unified. In addition, we now obtain two very short periods of two elements, rather than just one as seen in the usual periodic tables. So rather than having just one anomalous period length that does not repeat, the left-step table features all period lengths repeating once to give a sequence of 2, 2, 8, 8, 18, 18, etc. None of these advantages involve quantum mechanics but are features that Janet, who was not aware of this theory, appreciated. As we saw in Chapter 8, the introduction of quantum mechanics to the periodic table resulted in an understanding based upon electronic configurations. In this approach, the elements in the periodic table differ from each other according to the type of orbital occupied by the differentiating electron.

In the conventional tables, the elements in the two leftmost groups are said to form the s-block because their differentiating electrons enter an s-orbital. Moving towards the right, we encounter the d-block, then the p-block, and finally the f-block, the last of which lurks underneath the main body of the table. This order of blocks from left to right is not the most 'natural' or expected one, since the order of increasing energy of orbitals is:

$$s < p < d < f.$$

The left-step table maintains this sequence, although it is in reverse order. Whether this is really an advantage is debatable, however, since any cyclic table would also show this feature.

But there may be another advantage from the quantum mechanical point of view. There is no disputing the fact that the electronic configuration of the helium atom shows two electrons, both of which are in a 1s orbital. This should make helium an s-block element. But in conventional periodic tables, helium is placed among the noble gases because of its chemical properties, being highly inert just like the remaining noble gases (Ne, Ar, Kr, Xe, Rn).

This situation seems to provide a parallel with the earlier discussed historical case of the tellurium–iodine reversal in which atomic weight ordering had to be ignored in order to preserve chemical similarities. Similarly, in the case of helium there appear to be two possibilities:

1. Electronic structure is not the final arbiter over the placement of elements into groups and may well be replaced by some new criterion in due course (for example, atomic weight was eventually replaced by atomic number in the ordering of the elements, thus solving the pair reversal problem).
2. We do not really have a parallel case and electronic configuration still rules the day, in which case the apparently inert chemistry of helium should be ignored.

Notice that option 1 actually favours the conventional periodic tables, whereas option 2 favours the left-step table. Clearly, it is not so easy to decide whether the left-step table presents an advantage from the quantum mechanical point of view. Let me now throw another idea into the mix. Recall what was said in Chapter 4 on the nature of the elements and how Mendeleev, in particular, favoured regarding elements in the more abstract sense, rather than being tied to the elements as being simple or isolated substances. This reliance on the abstract sense of an element could be used to justify moving helium to the alkaline earth group. The concern that helium's chemical unreactivity denies it a place among the more reactive alkaline earths can thus be countered by putting one's attention on the nature of the element as an abstract entity rather than on its chemical properties. However, this move amounts to saying 'why not' place helium in the alkaline earths if its chemistry can be ignored?

I end this section by reporting an intriguing experiment that was conducted in 2017, in which Artem Oganov led a team that produced a genuine compound of helium and sodium, Na_2He, admittedly at an extremely high pressure. The existence of this apparently genuine compound of helium suggests that neon, rather than helium, is the least reactive of the noble gases, and has reopened the debate on whether the left-step periodic table might represent the optimal table after all.

Atomic number triads applied to group 3

There is a long-standing dispute among chemists and chemical educators regarding group 3 of the periodic table. Some old periodic tables show the following elements in group 3:

Sc
Y
La
Ac

More recently, periodic tables shown in many textbooks began to show group 3 as:

Sc
Y
Lu
Lr

with arguments based on presumed electronic configurations. In 1986, an article was published by William Jensen of the University of Cincinnati, in which he argued that textbook authors and periodic table designers should show group 3 as: Sc, Y, Lu, and Lr.

Then even more recently, a rearguard action occurred in which some authors have argued for a return to Sc, Y, La, Ac. How, if at all, does the notion of atomic number triads bear on this group 3 issue? If we consider atomic number triads, the answer is once again categorical, namely that Jensen's grouping is favoured. Whereas the first triad below is exact,

Y	39
Lu	$71 = (39 + 103)/2$
Lr	103

the second triad is incorrect,

Y	39
La	$57 \neq (39 + 89)/2 = 64$
Ac	89

But there is another reason why Jensen's arrangement is the better one, and this does not depend on any allegiance to atomic number triads.

If we consider the long-form periodic table and attempt to incorporate either lutetium and lawrencium or lanthanum and actinium into group 3, only the first placement makes any sense because it results in a continuously increasing sequence of atomic numbers. Conversely, the incorporation of lanthanum and actinium into group 3 of a long-form table results in two rather glaring anomalies in terms of sequences of increasing atomic numbers (Figure 40).

Finally, there is actually a third possibility, but this involves a rather awkward subdivision of the d-block elements as shown in Figure 41. Although some books include a periodic table arranged like Figure 41, this is not a very popular design and for rather obvious reasons. Presenting the periodic table in this fashion requires that the d-block of the periodic table should be split up into two highly unequal portions containing a block that is only one element wide and another block containing a nine-element-wide block. Given that this behaviour does not occur in any other block in the periodic table, it would appear to be the least likely of the three possible tables to reflect the actual arrangement of the elements in nature.

It should be acknowledged that there is in fact one block of the traditional periodic table, the s-block, whose first row is generally split into two portions, in that hydrogen and helium are separated in the traditional eighteen- or thirty-two-column tables. Nevertheless, this separation is carried out in a more plausible symmetrical fashion, and moreover is avoided altogether in the Janet left-step table as we have seen.

H 1																															He 2
Li 3	Be 4																									B 5	C 6	N 7	O 8	F 9	Ne 10
Na 11	Mg 12																									Al 13	Si 14	P 15	S 16	Cl 17	Ar 18
K 19	Ca 20															Sc 21	Ti 22	V 23	Cr 24	Mn 25	Fe 26	Co 27	Ni 28	Cu 29	Zn 30	Ga 31	Ge 32	As 33	Se 34	Br 35	Kr 36
Rb 37	Sr 38															Y 39	Zr 40	Nb 41	Mo 42	Tc 43	Ru 44	Rh 45	Pd 46	Ag 47	Cd 48	In 49	Sn 50	Sb 51	Te 52	I 53	Xe 54
Cs 55	Ba 56	La 57	Ce 58	Pr 59	Nd 60	Pm 61	Sm 62	Eu 63	Gd 64	Tb 65	Dy 66	Ho 67	Er 68	Tm 69	Yb 70	*Lu* 71	Hf 72	Ta 73	W 74	Re 75	Os 76	Ir 77	Pt 78	Au 79	Hg 80	Tl 81	Pb 82	Bi 83	Po 84	At 85	Rn 86
Fr 87	Ra 88	Ac 89	Th 90	Pa 91	U 92	Np 93	Pu 94	Am 95	Cm 96	Bk 97	Cf 98	Es 99	Fm 100	Md 101	No 102	*Lr* 103	Rf 104	Db 105	Sg 106	Bh 107	Hs 108	Mt 109	Ds 110	Rg 111	Cn 112	Nh 113	Fl 114	Mc 115	Lv 116	Ts 117	Og 118

H 1																															He 2
Li 3	Be 4																									B 5	C 6	N 7	O 8	F 9	Ne 10
Na 11	Mg 12																									Al 13	Si 14	P 15	S 16	Cl 17	Ar 18
K 19	Ca 20															Sc 21	Ti 22	V 23	Cr 24	Mn 25	Fe 26	Co 27	Ni 28	Cu 29	Zn 30	Ga 31	Ge 32	As 33	Se 34	Br 35	Kr 36
Rb 37	Sr 38															Y 39	Zr 40	Nb 41	Mo 42	Tc 43	Ru 44	Rh 45	Pd 46	Ag 47	Cd 48	In 49	Sn 50	Sb 51	Te 52	I 53	Xe 54
Cs 55	Ba 56	Ce 58	Pr 59	Nd 60	Pm 61	Sm 62	Eu 63	Gd 64	Tb 65	Dy 66	Ho 67	Er 68	Tm 69	Yb 70	Lu 71	*La* 57	Hf 72	Ta 73	W 74	Re 75	Os 76	Ir 77	Pt 78	Au 79	Hg 80	Tl 81	Pb 82	Bi 83	Po 84	At 85	Rn 86
Fr 87	Ra 88	Th 90	Pa 91	U 92	Np 93	Pu 94	Am 95	Cm 96	Bk 97	Cf 98	Es 99	Fm 100	Md 101	No 102	Lr 103	*Ac* 89	Rf 104	Db 105	Sg 106	Bh 107	Hs 108	Mt 109	Ds 110	Rg 111	Cn 112	Nh 113	Fl 114	Mc 115	Lv 116	Ts 117	Og 118

40. Long-form periodic tables showing alternative placements of Lu and Lr. Only the upper version preserves a continuous sequence of atomic numbers.

H																															He
Li	Be																									B	C	N	O	F	Ne
Na	Mg																									Al	Si	P	S	Cl	Ar
K	Ca	Sc															Ti	V	Cr	Mn	Fe	Co	Ni	Cu	Zn	Ga	Ge	As	Se	Br	Kr
Rb	Sr	Y															Zr	Nb	Mo	Tc	Ru	Rh	Pd	Ag	Cd	In	Sn	Sb	Te	I	Xe
Cs	Ba	La	Ce	Pr	Nd	Pm	Sm	Eu	Gd	Tb	Dy	Ho	Er	Tm	Yb	Lu	Hf	Ta	W	Re	Os	Ir	Pt	Au	Hg	Tl	Pb	Bi	Po	At	Rn
Fr	Ra	Ac	Th	Pa	U	Np	Pu	Am	Cm	Bk	Cf	Es	Fm	Md	No	Lr	Rf	Db	Sg	Bh	Hs	Mt	Ds	Rg	Cn	Nh	Fl	Mc	Lv	Ts	Og

41. Third option for presenting long-form periodic table in which the d-block is split into two uneven portions of one and nine groups.

The case of hydrogen and atomic number triads reconsidered

Another element which has traditionally caused problems is the very first one, namely hydrogen. In chemical terms, it seems to belong to group 1 (alkali metals) due to its ability to form +1 ions of H^+. Alternatively, H is rather unique in that it can also form an H^- ion, as in the case of metal hydrides such as NaH and CaH_2. This behaviour supports the placement of hydrogen in group 17 (the halogens) which also form –1 ions. How can this issue be settled categorically? Some authors take an easy way out by allowing hydrogen to 'float' majestically above the main body of the periodic table; in other words, they do not commit themselves to either one of the possible placements.

This has always seemed to me a display of 'chemical elitism', since it seems to imply that whereas all the elements are subject to the periodic law, hydrogen is somehow a special case and as such is above the law, very much like the British royal family once was. In the first edition of this book and until quite recently I had proposed considering the formation of a new atomic number triad in order to settle the question of the placement of hydrogen. If one uses this approach it produces a very clear-cut result in favour of placing hydrogen among the halogens rather than the alkali metals. In the conventional periodic table, in which hydrogen is placed in the alkali metals, there is no perfect triad, whereas if hydrogen is allowed to sit at the top of the halogens, a new atomic number triad comes into being (Figure 42):

H	1		H	1	
Li	3	$(1+11)/2 \neq 3$	F	9	$(1+17)/2=9$
Na	11		Cl	17	

However appealing this proposal might seem, I now think that it may represent a mistaken strategy. My reason for saying so is that

																												H	He	Li	Be
																								B	C	N	O	F	Ne	Na	Mg
																								Al	Si	P	S	Cl	Ar	K	Ca
														Sc	Ti	V	Cr	Mn	Fe	Co	Ni	Cu	Zn	Ga	Ge	As	Se	Br	Kr	Rb	Sr
														Y	Zr	Nb	Mo	Tc	Ru	Rh	Pd	Ag	Cd	In	Sn	Sb	Te	I	Xe	Cs	Ba
La	Ce	Pr	Nd	Pm	Sm	Eu	Gd	Tb	Dy	Ho	Er	Tm	Yb	Lu	Hf	Ta	W	Re	Os	Ir	Pt	Au	Hg	Tl	Pb	Bi	Po	At	Rn	Fr	Ra
Ac	Th	Pa	U	Np	Pu	Am	Cm	Bk	Cf	Es	Fm	Md	No	Lr	Rf	Db	Sg	Bh	Hs	Mt	Ds	Rg	Cn	Nh	Fl	Mc	Lv	Ts	Og	119	120

42. Periodic table based on maximizing atomic number triads.

the first members of groups of elements are never members of triads and there is no reason to believe that a group such as the halogens should represent an exception.

It is interesting to examine the conditions under which correct triads of elements occur in the periodic table. In the case of conventional periodic tables that feature the s-block on the left edge of the table an anomaly can be seen. For s-block elements alone, an atomic number triad of elements only occurs if the first and second elements occur in periods of equal lengths as in the case of lithium (3), sodium (11), and potassium (19). However, in the case of the p- and d-blocks, and in principle even the f-block, triads of elements only occur if the second and third elements fall into periods of equal lengths, such as chlorine (17), bromine (35), and iodine (53).

But if the elements are presented in the form of a left-step table all triads of elements without exception consist of sets of three elements in which the second and third of them belong to periods of equal lengths (Figure 43). I believe that this may present a further, although admittedly formal, argument in favour of the superiority of the left-step periodic table. It is also my belief that this table may perhaps be the long-sought optimal periodic table.

Is there an official IUPAC periodic table?

The International Union for Pure and Applied Chemistry, the governing body for chemical nomenclature, maintains a rather ambiguous position regarding the best form of the periodic table. Even though the official policy of IUPAC is that they do not recommend any particular form, their literature often features a periodic table shown in Figure 44 which contains two fifteen-element-wide series of lanthanoids and actinoids. The disadvantage of such a format is that it leaves group 3 of the table with only two elements, something that does not occur in any

																														1	2
																														3	4
																								5	6	7	8	9	10	11	12
																								13	14	15	16	17	18	19	20
														21	22	23	24	25	26	27	28	29	30	31	32	33	34	35	36	37	38
														39	40	41	42	43	44	45	46	47	48	49	50	51	52	53	54	55	56
57	58	59	60	61	62	63	64	65	66	67	68	69	70	71	72	73	74	75	76	77	78	79	80	81	82	83	84	85	86	87	88
89	90	91	92	93	94	95	96	97	98	99	100	101	102	103	104	105	106	107	108	109	110	111	112	113	114	115	116	117	118	119	120

43. Atomic number triads highlighted on a left-step periodic table. All triads contain their second and third elements in periods of equal lengths.

Group #

1	2	3	4	5	6	7	8	9	10	11	12	13	14	15	16	17	18
H																	He
Li	Be											B	C	N	O	F	Ne
Na	Mg											Al	Si	P	S	Cl	Ar
K	Ca	Sc	Ti	V	Cr	Mn	Fe	Co	Ni	Cu	Zn	Ga	Ge	As	Se	Br	Kr
Rb	Sr	Y	Zr	Nb	Mo	Tc	Ru	Rh	Pd	Ag	Cd	In	Sn	Sb	Te	I	Xe
Cs	Ba		Hf	Ta	W	Re	Os	Ir	Pt	Au	Hg	Tl	Pb	Bi	Po	At	Rn
Fr	Ra		Rf	Db	Sg	Bh	Hs	Mt	Ds	Rg	Cn	Nh	Fl	Mc	Lv	Ts	Og

La	Ce	Pr	Nd	Pm	Sm	Eu	Gd	Tb	Dy	Ho	Er	Tm	Yb	Lu
Ac	Th	Pa	U	Np	Pu	Am	Cm	Bk	Cf	Es	Fm	Md	No	Lr

44. IUPAC periodic table with fifteen-element-wide lanthanoid and actinoid blocks. Group 3 contains only two elements.

other group whatsoever. The anomaly is heightened by displaying this kind of periodic table in a thirty-two-column format as shown in Figure 45. I believe it might be time for IUPAC to recognize the problems with such a display and to make an official ruling on the most consistent currently available table, which, as I have suggested, consists of the upper version shown in Figure 40.

I hope that the reader comes away from this book with the realization that even 150 years after the periodic table was formulated, it is still very much a subject of interest and development.

H																																He
Li	Be																										B	C	N	O	F	Ne
Na	Mg																										Al	Si	P	S	Cl	Ar
K	Ca																Sc	Ti	V	Cr	Mn	Fe	Co	Ni	Cu	Zn	Ga	Ge	As	Se	Br	Kr
Rb	Sr																Y	Zr	Nb	Mo	Tc	Ru	Rh	Pd	Ag	Cd	In	Sn	Sb	Te	I	Xe
Cs	Ba	La	Ce	Pr	Nd	Pm	Sm	Eu	Gd	Tb	Dy	Ho	Er	Tm	Yb	Lu		Hf	Ta	W	Re	Os	Ir	Pt	Au	Hg	Tl	Pb	Bi	Po	At	Rn
Fr	Ra	Ac	Th	Pa	U	Np	Pu	Am	Cm	Bk	Cf	Es	Fm	Md	No	Lr		Rf	Db	Sg	Bh	Hs	Mt	Ds	Rg	Cn	Nh	Fl	Mc	Lv	Ts	Og

45. A thirty-two-column version that would be consistent with the table published in IUPAC literature, in which the lanthanoids and actinoids are presented as fifteen-element-wide series.

Further reading

M. Fontani, M. Costa, and M.V. Orna, *The Lost Elements*, Oxford University Press, 2015.
M. Gordin, *A Well-Ordered Thing*, Princeton University Press, 2018.
M. Kaji, H. Kragh, and G. Pallo, *Early Responses to the Periodic Table*, Oxford University Press, 2015.
S. Kean, *The Disappearing Spoon: And Other True Tales of Madness, Love, and the History of the World from the Periodic Table of the Elements*, Little, Brown and Company, 2010.
E. Mazurs, *Graphical Representations of the Periodic System during 100 Years*, University of Alabama Press, 1974.
F. A. Paneth, The epistemological status of the chemical concept of element, *British Journal for the Philosophy of Science*, 13, 1–14, 144–60, 1962.
E. R. Scerri, *The Periodic Table, Its Story and Its Significance*, Oxford University Press, 2007.
E. R. Scerri, The trouble with the Aufbau principle, *Education in Chemistry*, 24–6, November, 2013.
E. R. Scerri, *A Tale of Seven Elements*, Oxford University Press, 2013.
E. R. Scerri, The changing views of a philosopher of chemistry on the question of reduction, in E. Scerri and G. Fisher, *Essays in The Philosophy of Chemistry*, Oxford University Press, 2016, pp. 125–43.
E. R. Scerri, Can quantum ideas explain chemistry's greatest icon? *Nature*, 565, 557–9, 2019.
E. R. Scerri and W. Parsons, What elements belong in group 3? In E. R. Scerri and G. Restrepo, eds, *Mendeleev to Oganesson*, Oxford University Press, 2018, pp. 140–51.

W. H. E. Schwarz et al., *Chemistry—A European Journal*, 12, 4101, 2006 <https://doi.org/10.1002/chem.200500945> (see Figure 8 and accompanying text).

P. Thyssen and A. Ceulemans, *Shattered Symmetry*, Oxford University Press, 2017.

J. W. van Spronsen, *The Periodic System of Chemical Elements: A History of the First Hundred Years*, Elsevier, 1969.

Recommended websites

Mark Leach's metasynthesis site. A wonderful compendium of periodic tables: <http://www.meta-synthesis.com/webbook/35_pt/pt_database.php>.

Mark Winters' Webelements site: <https://www.webelements.com/>.

Eric Scerri's website for history and philosophy of chemistry and the periodic table: <http://ericscerri.com/>.

“牛津通识读本”已出书目

古典哲学的趣味
人生的意义
文学理论入门
大众经济学
历史之源
设计，无处不在
生活中的心理学
政治的历史与边界
哲学的思与惑
资本主义
美国总统制
海德格尔
我们时代的伦理学
卡夫卡是谁
考古学的过去与未来
天文学简史
社会学的意识
康德
尼采
亚里士多德的世界
西方艺术新论
全球化面面观
简明逻辑学
法哲学：价值与事实
政治哲学与幸福根基
选择理论
后殖民主义与世界格局

福柯
缤纷的语言学
达达和超现实主义
佛学概论
维特根斯坦与哲学
科学哲学
印度哲学祛魅
克尔凯郭尔
科学革命
广告
数学
叔本华
笛卡尔
基督教神学
犹太人与犹太教
现代日本
罗兰·巴特
马基雅维里
全球经济史
进化
性存在
量子理论
牛顿新传
国际移民
哈贝马斯
医学伦理
黑格尔

地球
记忆
法律
中国文学
托克维尔
休谟
分子
法国大革命
民族主义
科幻作品
罗素
美国政党与选举
美国最高法院
纪录片
大萧条与罗斯福新政
领导力
无神论
罗马共和国
美国国会
民主
英格兰文学
现代主义
网络
自闭症
德里达
浪漫主义
批判理论

德国文学
戏剧
腐败
医事法
癌症
植物
法语文学
微观经济学
湖泊
拜占庭
司法心理学
发展
农业
特洛伊战争
巴比伦尼亚
河流
战争与技术
品牌学
数学简史
儿童心理学
时装
现代拉丁美洲文学
卢梭
隐私
电影音乐
抑郁症
传染病
希腊化时代
知识
环境伦理学
美国革命
元素周期表
人口学
社会心理学
动物
项目管理
美学
管理学
电影
俄罗斯文学
古典文学
大数据
洛克
幸福
免疫系统
银行学
景观设计学
神圣罗马帝国
大流行病
亚历山大大帝
气候
第二次世界大战
中世纪
工业革命
传记
公共管理
社会语言学